Robert E. Gough, PhD

The Highbush Blueberry and Its Management

*Pre-publication
REVIEWS,
COMMENTARIES,
EVALUATIONS . . .*

More pre-publication

REVIEWS, COMMENTARIES, EVALUATIONS . . .

"**V**ery informative. Everything I read was down to earth . . . . The book is easy to read and understand without having to decipher the technical terms that other blueberry books have. I know all of our customers and blueberry growers will enjoy reading the book and will be more knowledgeable after reading it. THIS IS THE BEST BOOK I'VE SEEN FOR BLUEBERRIES."

Daniel P. Hartmann
Vice President/General Manager,
Hartmann's Plantation, Inc.,
Grand Junction, Michigan

"**T**his new book written by Dr. Bob Gough is AN EXCELLENT SOURCE OF INFORMATION ON BLUEBERRIES. Dr. Gough's writing is very comprehensible and all of the standard practical practices are written about our crop. THE BOOK IS WORTH ITS WEIGHT IN GOLD."

Northern & Southern
Grower's Guide

Food Products Press
An Imprint of The Haworth Press, Inc.

The Highbush Blueberry and Its Management

FOOD PRODUCTS PRESS
An Imprint of The Haworth Press, Inc.
Robert E. Gough, PhD, Senior Editor

New, Recent, and Forthcoming Titles:

The Highbush Blueberry and Its Management by Robert E. Gough

Glossary of Vital Terms for the Home Gardener by Robert E. Gough

Seed Quality: Basic Mechanisms and Agricultural Implications edited by Amarjit S. Basra

Statistical Methods for Food and Agriculture edited by Filmore E. Bender, Larry W. Douglass, and Amihud Kramer

World Food and You by Nan Unklesbay

Introduction to the General Principles of Aquaculture by Hans Ackefors, Jay V. Huner, and Mark Konikoff

Managing the Potato Production System by Bill B. Dean

Marketing Livestock and Meat by William Lesser

The World Apple Market by A. Desmond O'Rourke

Understanding the Japanese Food and Agrimarket: A Multifaceted Opportunity edited by A. Desmond O'Rourke

Marketing Beef in Japan by William A. Kerr et al.

The Highbush Blueberry and Its Management

Robert E. Gough, PhD

Food Products Press
An Imprint of The Haworth Press, Inc.
New York • London • Norwood (Australia)

Published by

Food Products Press, an imprint of The Haworth Press, Inc., 10 Alice Street, Binghamton, NY 13904-1580

Library of Congress Cataloging-in-Publication Data

Gough, Robert E. (Robert Edward)
    The highbush blueberry and its management / Robert E. Gough.
      p. cm.
    Includes bibliographical references and index.
    ISBN 1-56022-022-8 (alk. paper).
      1. Blueberries. I. Title.
SB386.B7G68 1991
634'.737–dc20
                              ISBN 1-56022-021-X (alk. paper).   92-225
                                                            CIP

To my family
and
to all students of highbush blueberry culture
the world over

ABOUT THE AUTHOR

Robert E. Gough, PhD, was formerly Associate Professor of Horticulture at The University of Rhode Island and President of the Northeast Region of the American Society for Horticultural Science. He is a leading specialist in small fruit and viticulture. Dr. Gough has a diverse background, having earned a BA in English, an MS in horticulture, and a PhD in botany. His experiences as a county agricultural agent, state and regional extension specialist in small fruit, and a senior research scientist have provided him with a great deal of insight into the needs of growers.

Dr. Gough has published extensively–nearly 300 articles–in the area of pomology (fruit science) in both scientific journals and popular magazines, including the *Journal of the American Society for Horticultural Science, HortScience, Scientia Horticulturae, Journal of Small Fruit & Viticulture, New England Gardener, Harrowsmith/Country Life, Country Journal, Fine Gardening, National Gardening, New England Farmer, American Fruit Grower, Proceedings of the New England Small Fruit Conference,* and *Proceedings of the Ohio Fruit Congress.* He also maintains an active interest in all areas relating to general horticulture, agriculture, crop science, soil science, and community gardening/landscape architecture. A member of numerous professional and honorary societies, Dr. Gough served as an associate editor for the *Journal of the American Society for Horticulture Science* and *HortScience* from 1985-1988. Currently, he is Special Projects Manager and Associate Editor of The Haworth Press, Inc., and General Editor of the Food Products Press imprint.

CONTENTS

Preface

This book is for the grower. It is a culmination of 20 years of research and of helping growers to produce quality crops of highbush blueberry. I hope it helps answer some of the questions posed to me by homeowners, small diversified growers, and larger specialized commercial growers during hundreds of twilight meetings, conferences, seminars, schools, and workshops. Parts of the book will be of limited interest to some, but they are necessary for a balanced treatment of the subject. Much of the information given is supported by references for those who wish to delve further into the exciting world of blueberry research. Some older references have been included because the findings are still relevant and many are considered classic studies. Newer references, citing work that is relevant to the grower, make up the bulk of the material. This book is focused on North American production practices but should provide good background material for growers throughout the world.

It would be nearly impossible to give the names of all the people who have contributed to this work over the years. My heartfelt thanks go to the following researchers who have offered suggestions and/or supplied information on highbush blueberry culture in their particular geographic areas: Dr. John Abbott, Ciba-Geigy, Lancaster, Pennsylvania; Professor Ramon Amenabar Arzuaga, Responsable Estacion Fruiticultura Zalla, Bilbao, Spain; Dr. G. Bunnemann, Institute fur Obstbau und Baumschule, Universitat Hannover, Sarstedt, Germany; Dr. N. F. Childers, University of Florida, Gainesville, Florida; Dr. John Clark, University of Arkansas, Fayetteville, Arkansas; Professor Kevin Clayton-Greene, Horticultural Research Institute, Victoria, Australia; Dr. A. D. Draper, United States Department of Agriculture, Beltsville, Maryland; Dr. David Handley, University of Maine, Orono, Maine; Dr. Paul Lyrene, University of Florida, Gainesville, Florida; Dr. C. M. Mainland, North Carolina State University, Raleigh, North Carolina; Professor Carlos E. Munoz S., Instituto de Investigaciones Agropecuarias Estacion Experimental La Platina, Santiago, Chile; Dr. John Nelson, Michigan Blueberry Growers, Grand Junction, Michigan; Dr. Gary Pavlis, Rutgers University, New Brunswick, New Jersey; Dr. Marvin Pritts, Cornell University, Ithaca, New York; Dr. James Spiers, United States Department of Agriculture, Poplarville, Mississippi; Dr. Takato Tamada, Chiba-ken Agricultural College, Chiba-ken, Japan; and Dr. J. M. Wijsmuller, Stichting Fruiteeltproeftuin voor Limburg en Noord Brabant, Horst, Holland.

A special thanks is also extended to the following growers and researchers who have offered their critical review of portions of the manuscript over the several years of its preparation: Dr. Walter Ballinger, North Carolina State University, Raleigh, North Carolina; Dr. Bertie Boyce, University of Vermont, Burlington, Vermont; Dr. Paul Eck, Rutgers University, New Brunswick, New Jersey; Dr. Richard Funt, Ohio State University, Columbus, Ohio; Dr. James Hancock, Michigan State University, East Lansing, Michigan; Mr. Dan Hartmann, Hartmann's Plantation, Grand Junction, Michigan; Dr. Amr Ismail, University of Maine, Orono, Maine; Dr. Ron Korcak, United States Department of Agriculture, Beltsville, Maryland; Dr. Michel Lareau, Agriculture Canada Research Station, Quebec, Canada; Professor William Lord, University of New Hampshire, Durham, New Hampshire; Dr. Paul Lyrene, University of Florida, Gainesville, Florida; Dr. Lloyd Martin, Oregon State University, Corvallis, Oregon; Dr. John McGuire, University of Rhode Island, Kingston, Rhode Island; Dr. James Moore, University of Arkansas, Fayetteville, Arkansas; Dr. Kazimierz Pliszka, Warsaw Agricultural University, Warsaw, Poland; Dr. J. Thompson, Cornell University, Ithaca, New York; and Professor D. B. Wallace, University of Rhode Island, Kingston, Rhode Island. A very special thanks goes to Dr. Vladimir Shutak, University of Rhode Island, Kingston, Rhode Island, who has been involved with this project since its inception and who has offered years of encouragement and professional help as teacher, mentor, and friend.

Without the kind help of all these people, the completion of this project would have been impossible. Since none have reviewed the entire manuscript, however, any shortcomings are solely the author's responsibility.

Chapter 1

Introduction

There are an estimated 350,000 species of plants in the world, of which only 1500 or so are of economic importance. Of these, about 150 are of major importance, but only 15 provide most of the world's food. The banana and coconut are the only fruit in the top-15 species. The other fruit, including the blueberry, fall somewhere in the top 15-1500 in importance as food.

Naming plants has gone on for centuries, but often only in an individual and highly unsystematic way. Theophrastus, the fifth-century B.C. Greek physician, began a cumbersome systematic classification identifying each plant with long descriptions. This unwieldy procedure remained in place for 20 centuries. The Swedish scientist Carl von Linne (1707-1778), better known as Carolus Linnaeus, or simply Linnaeus, devised a better system. Each plant was assigned to several plant groups, each more specific and exclusive than the former. The groups were assigned classical Latin or Greek names, since those languages were dead and would remain static indefinitely. Linnaeus believed that modern languages, on the other hand, were dynamic, changing constantly through use, so plant descriptions employing them would be rendered obsolete in a short time. In addition, the classical languages were universally understood by scholars throughout the world.

Linnaeus grouped all plants into the plant kingdom Plantae, to separate them from animals. The blueberry is further classed with all plants producing flowers and seeds into the division Spermatophyta. The flower has an ovary, thus placing it in the class Angiospermae, and its seedling has two leaves, putting it in the subclass Dicotyledonae. It is a member of the Ericaceae family of plants comprising mostly woody shrubs that grow naturally on acid soils. This is a large family and is found widely distributed throughout the world. It includes, among others, the rhododendrons, azaleas, heathers, heaths, and mountain laurels. The blueberry belongs to the subfamily Vacciniaceae; the tribe Vaccinieae; the genus *Vaccinium*; and the subgenus *Cyanococcus* (from the Greek *cyano* ["blue"] and *coccus* ["berry"]). There are many species. The terms "genus" (pl. genera) and "species" are most commonly used when discussing plants horticulturally. The first letter of the genus name is always capitalized; the first letter of the species name is lower-case. Both are italicized or underlined. Among growers, species are rarely mentioned, but a more specific term, "cultivar," is common. This term–a contraction of the words "cultivated variety"–is always written in a modern language, usually English, and is preceded by the word "cultivar," its abbreviation (cv.), or placed in single quotes.

Members of the blueberry genus, the name of which means "hyacinth" in classical Latin, became highly differentiated and developed prior to the Cretaceous period more than 100 million years ago. Forms of blueberry plants that had adapted to temperate climates evolved from tropical forms and became firmly established– predominantly in eastern North America–after the Pleistocene glaciations (or ice sheets), the latest of which receded some 12,000 years ago. Plants of the genus *Vaccinium*, section *Cyanococcus* hybridized freely in the wild and moved rapidly into areas disturbed by ice sheets and fires. Wild animals disseminated the seeds of some species in their droppings, while the plants themselves spread by underground runners, or rhizomes (Camp 1945).

The highbush blueberry ranges from 5-23 ft (1.5-7 m) in height. The cultivated highbush blueberry was developed primarily from two species: *V. corymbosum* L. and *V. australe* Small, though other species have been utilized in modern breeding programs. The letter, name, or abbreviated name following the species refers to the person who first named it (e.g., "L." is the abbreviation for Linnaeus). Wild plants are distributed in sunny, acidic, and swampy areas from Nova Scotia and southern Quebec west to Wisconsin and south to northern Florida and southeastern Alabama. Wild southern populations are comprised primarily of *V. australe* Small, whereas *V. corymbosum* L. occurs in more northerly areas. Because of rampant hybridization, these species have intermingled and crossed with a half-dozen other economically minor species, thus giving rise to various intermediate forms.

Although dealing with scientific names can at times be trying, dealing with common names is worse, for people in different parts of the world use different names for the same plant. Often confused are the blueberries and the huckleberries, each name having been applied to species in the other group at some time. Formerly, both were given the same genus, *Vaccinium*, but in the mid-19th century, they were placed into separate genera (Darlington 1847). Those plants having small, resinous dots (resembling spattered varnish) on the undersides of the leaves and having fruit with ten large, bony seeds were placed in a new genus, *Gaylussacia*, named for the French chemist Gay Lussac. These plants were given the common name of huckleberry. The other species remained in the genus *Vaccinium* and were called blueberry. Unfortunately, old customs die hard, so the common names are still often used interchangeably. For example, the bilberry, whortleberry, mortinia, blaeberry, whinberry, red huckleberry, deerberry, squawberry, moorberry, cowberry, American cranberry, foxberry, blue huckleberry, tall huckleberry, swamp huckleberry, coast huckleberry, low huckleberry, hairy huckleberry, and Mayberry are all members of *Vaccinium*. Members of *Gaylussacia* are commonly called huckleberries, with no modifying adjective. The highbush blueberry, *V. corymbosum* L.–also called blue huckleberry, tall huckleberry, swamp huckleberry, high blueberry, and swamp blueberry–is the progenitor of the modern blueberry industry and is the subject of this work. Even in the mid-19th century, there was no fruit of any species so admired on the New York and Philadelphia markets as that of the highbush blueberry (Darlington 1847).

Many species of blueberries were plentiful before and during the European colonization of North America, and these were held in high esteem by the natives. According to legend, the Great Spirit sent the "star berries" to relieve famine. The berries were so named because of their star-shaped calyx (Ascher 1991). Anne Pollard, the 12-year-old Puritan who was the first ashore after the 1630 landing, wrote that Boston's Beacon Hill was covered with blueberries (Russell 1980). Josselyn, a New England traveler of the early 1600s, called the numerous "Skycoloured" berries a most excellent summer dish that the colonists ate in milk and sweetened with sugar and spice (Russell 1982). Kalm, Champlain, and other early travelers into the hinterlands of America recorded that the colonists learned from the natives to sun-dry the fruit for winter use in puddings, cakes, bread, and pemmican (Hedrick 1919; Russell 1980). As a display of generosity and honor, the Iroquois offered fresh blueberry corn bread to the white settlers (Benson 1966). Rather than relying upon uncertain sunlight, natives along the foggy, rainy Northwest Pacific coast smoke-dried their blueberries. Today's fruitcakes and breads are direct descendants of those native favorites. Rhode Island founder Roger Williams (1643) described similar use of these fruit, reporting that the dried fruit, when powdered and mixed with parched meal, made a very delicate dish called "sautauthig," which resembled spice cake. Other accounts indicate that some tribes added ducks and other game to stewed blueberries. In addition to eating the fresh berries, the natives also feasted upon the great flocks of passenger pigeons lighting on the bushes in late summer (Cronon 1983).

Fruit of related species are utilized around the world (Hedrick 1919). Spaniards enjoy their native, black, juicy, Maderia whortleberry. In Jamaica, the sour, red Jamaica bilberry is widely used to make a jelly. The berries of mortima appear on local markets in Ecuador and Colombia. The Highlanders of Scotland eat their blaeberries in milk or tarts. Natives of the off-coast Orkneys make a wine from theirs, which belong to the same species the Rocky Mountain natives enjoyed. In Siberia, the bog bilberry was fermented and distilled into a strong drink, while in France it was used at one time to color wine. While enjoying the Maine woods, Thoreau stewed and sweetened the cowberry for dessert. These fruit, also known as the mountain cranberry, or lingonberry, are highly valued for jellies in modern Scandinavia.

Other parts of the highbush blueberry plant were also utilized by both colonists and natives. The leaves, when chewed, yield a drug known as vaccinium; a tea made from the leaves and fruit was a remedy for diarrhea and suppressed menstruation (Russell 1980). Infusions of the flowers and rhizomes were used to treat infant colic, to induce labor, and to purify the blood (Vander Kloet 1988). The strong, flexible wood made excellent tool handles.

Chapter 2

Selection and Improvement

In spite of its popularity with early settlers, the blueberry was among the most recent fruits to be domesticated. This was probably due to its abundance in the wild, making cultivation unnecessary (Card 1903). With the burgeoning U.S. population, demand for the fruit increased at the same time suitable habitat decreased, thus forcing the plant into cultivation around the turn of the 20th century. Some earlier attempts at transplanting wild plants were successful—notably those of *V. atrococcum*, the black highbush blueberry, on the grounds of the Smithsonian Institution around 1850—but most efforts failed because the plant's requirement for acid soil was unknown.

During the 1890s, various plant scientists in Maine, Michigan, New York, Rhode Island, and other areas made limited efforts to select and transplant particularly good wild bushes for commercial production. Again, none of these was particularly successful.

Recognizing both the potential widespread commercial value of the blueberry and the demand for the fruit on the Boston market, Dr. Frederick V. Coville, a botanist with the United States Department of Agriculture, began extensive research on the plant in 1906. He joined forces with Elizabeth C. White, a commercial cranberry grower in New Jersey, who had been instructing her workers to select and transplant especially good wild blueberry plants from the wetlands surrounding her bog. An early criterion for selection was large berry size, the standard gauge of which was Mrs. White's wedding band. The nagging problem of poor "take" of transplants was finally solved by Dr. Coville, who observed that bushes planted in relatively infertile, natural soils thrived, while those planted in rich, well-manured, and limed garden soils declined and died. Ex-

5

periments with potted plants in 1908 and 1909 confirmed that blueberries grow best in acid soils that are not highly manured and limed (Coville 1910). In this same classic work, Coville reported that the blueberry root is devoid of the root hairs that aid "ordinary" plants in water and nutrient absorption and, further, that the roots are inhabited by mycorrhizal fungus that seems to benefit the plants in unknown ways.

After discovering the plants' soil requirements, Dr. Coville devoted another two years to their culture from seed to fruit and investigated methods of propagating and pollinating the bushes. In 1908, the first wild highbush blueberry plant for breeding purposes was selected in Greenfield, New Hampshire, and named 'Brooks.' Its fruit were more than 0.5 in (1.3 cm) in diameter and highly flavored, a quality of primary importance to the success of the blueberry breeding program. Through failed attempts at self-pollination of plants in 1909 and 1910, Coville realized that cross-pollination resulted in the highest fruit production. A second wild plant, a lowbush blueberry, was selected for breeding in 1909, again from a pasture in Greenfield, New Hampshire. It was named 'Russell.' The first successful blueberry cross was made between 'Russell' and 'Brooks' in 1911. Subsequent crosses of their progeny in 1913 resulted in about 3000 hybrids, two of which were later released as 'Redskin' and 'Catawba.' In 1911, with Mrs. White's help, Coville selected 'Sooy' and seven other plants from the wilds around Browns Mills, New Jersey. A cross of 'Brooks' and 'Sooy' in 1912 resulted in another 3000 seedlings. Among these were noteworthy plants later released as 'Pioneer' and 'Katherine.' These two, together with another interspecific cross, 'Cabot,' and the wild selection 'Rubel,' formed the basis of the modern blueberry industry. By the time of his death in 1937, Dr. Coville had propagated over 68,000 seedlings, from which he had selected and introduced 15 improved cultivars (some of them third-generation hybrids). He introduced his last selection only a few months before his death and, ironically, named it 'Dixi,' which is Latin for "I am done" (Coville 1937).

Coville and others realized that interspecific crosses could readily be made between species with the same chromosome number (homoploids). He recorded successful hybrids between *Vaccinium sta-*

mineum L. (deerberry) and *V. myrtilloides* Mich. as well as between *V. melanocarpum* Mohr. and *V. myrtilloides* Mich. The discovery of interspecific hybridization permitted plant breeders to combine desirable traits (such as cold hardiness, higher sugar content, and drought tolerance) of several species into a single plant. For example, the extreme cold tolerance of *V. uliginosum* L., a bilberry, was introduced into the cultivated highbush 'Pemberton' by a Finnish plant breeder. The progeny of this union survives and produces high-quality fruit under very cold conditions of USDA hardiness zone 3, where average annual minimum temperatures reach −30° to −40°F (−34° to −40°C). In an analogous way, the genes of *V. altomontanum* Ashe and *V. membranaceum* Douglas contribute to drought resistance, whereas those of *V. darrowi* Camp provide low chilling characteristics. Dr. Coville, using 'Russell,' introduced lowbush blueberry genes into more than 15 highbush selections, an accomplishment that led to the development of "half-highs," which are plants 18-48 in (45-120 cm) high. Half-highs are extremely productive, early-ripening, and well-adapted to northern areas.

Interspecific hybridization ensures the diversification not only of the blueberry industry but also of the gene pool available to growers. This diversification is nature's way of guaranteeing that no single natural calamity will obliterate blueberry production. Until recently, genetic diversity among cultivated forms was limited, since most of the highbush blueberry cultivars could be traced back to the four plants Coville originally selected: 'Brooks,' 'Russell,' 'Sooy,' and 'Rubel' (Hancock and Siefker 1982). Through interspecific hybridizing, contemporary breeders have provided the industry with a "life insurance policy."

Increased genetic diversity and hardiness have not been the only breeding objectives of Coville's successors. Between 1939 and 1952, an additional 15 cultivars originating from Coville's stock were released. All were selected for other characteristics. Thanks to work on unravelling the inheritance patterns of some fruit characteristics (Darrow et al. 1939), researchers realized that 'Stanley' could pass on its large fruit size, 'Russell' its good scar and early ripening, and 'Grover' its dark color. In the meantime, other researchers uncovered the inheritance patterns of bush characteristics (Johnston 1942) and disease tolerance (Moore et al. 1962). Finn and Luby

(1992) reported that *V. angustifolium* parents produced progeny with dark, soft fruit with large scars.

The modern USDA breeding program is based at Chatsworth, New Jersey, and it works cooperatively with research stations and grower associations. Modern breeding objectives include a continued increase in berry quality and resistance to leading diseases, including mummy-berry and root rot. Further, new cultivars have been introduced that are adapted to non-traditional blueberry areas. Recently, researchers have extended the range of highbush production into the deep south, from the central Florida peninsula to coastal South Carolina and west along the Gulf Coast into southeast Texas. Northern highbush cultivars require more winter chilling than is available in this region. Such cultivars–when crossed with such southern species as the evergreen *V. darrowi* Camp., *V. elliottii* Chapm., *V. australe* Small, and *V. fuscatum* Ait.–produce offspring with the early fruit maturity of the northern highbush and the low chilling requirements of the southern blueberries (Patten et al. 1991). These offspring, in turn, can be quite productive under southern conditions (Sherman and Lyrene 1991). Such southern cultivars must have chilling requirements low enough to allow bloom but high enough to keep buds dormant during the often extended midwinter periods, when daily maximum temperatures approach 80°F (26.4°C). Unfortunately, some years may be too warm for adequate chilling, while in others, a cold December followed by a warm January can cause plants to flower several weeks prior to the average date of last frost. Variable soils, only about 10% of which are suitable for blueberries, and major problems with *Phytophthora* root rot and *Botrytis* flower blight may be alleviated by the future introduction of improved southern highbush cultivars (Lyrene, personal communication).

To extend the range of blueberry production into northern areas, breeders following Dr. Coville's lead have crossed the highbush with the lowbush species to reduce plant height, thus taking advantage of insulating snow cover, while at the same time increasing fruit size. Many of the progeny also have flexible canes that bend, but do not break, under snow. The half-highs are often twiggy and strongly rhizomatous, and they may spread out of their rows when planted too far south.

The gene pool for today's industry is expanding rapidly as new methods of interspecific hybridization are developed. In the future, higher-quality fruit will be grown over a wider environmental range than ever before. Many cultivars planted today bear fruit over five times larger than Coville's first selections and they are grown under conditions once thought unsuitable for production. In less than a century after the first wild plants were selected, well over 50 improved cultivars have been introduced. Almost all result from our newfound understanding of genetic manipulation and hybridization.

For a more technical discussion of highbush blueberry breeding techniques, refer to *Advances in Fruit Breeding* (Galletta 1975), *Blueberry Culture* (Eck and Childers 1966), and *Blueberry Science* (Eck 1988).

Chapter 3

Growth and Development

ANATOMY

Vegetative Anatomy

Roots

The highbush blueberry has two major types of roots: (1) pencil-thick storage and anchor roots and (2) fine, thread-like roots, often only 50 μm in diameter (Figure 3-1). The latter are the feeder roots responsible for nutrient absorption. The roots are composed of a thin root cap covering the apical meristem. The cap is somewhat muci-laginous and aids the root's penetration of the soil while protecting the meristematic tissue. The meristem tissue undergoes rapid cell division and is responsible in part for adding new cells to the root. The root structure 12-25 mm behind the apex shows a homogenous cortical region of 3-4 cell layers about 40 μm thick. The cells are simple parenchyma and alternate with one another in successive concentric layers. The cortex is surrounded by a poorly developed epidermal layer of symmetrical cells, the outer surfaces of which are convex. The cortex itself surrounds a single layer of cells called the endodermis, beneath which lies the vascular cylinder, composed of the vascular system and associated parenchyma. This inner cylinder, also called the stele, is about 30 μm in diameter. The blueberry root has no root hairs. The anatomy of older roots is similar to that of the stem.

Shoots

The shoot supports the buds and leaves of the plant and provides for conduction between the leaves and the roots. Older shoots are called canes.

Figure 3-1. A portion of the root system of a two-year-old plant. The thicker roots are for anchorage and storage; the thinner roots are feeder roots. A common pin is shown for comparison.

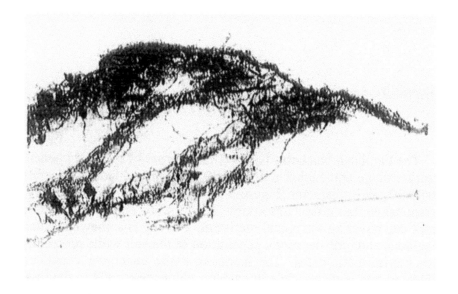

The blueberry shoot consists of a dense layer of ordered symmetrical epidermal cells surrounding up to a dozen layers of chlorophyllous cortical parenchyma tissue containing air canals. These are surrounded by isodiametric parenchyma cells attached to each other by thread-like constructions. Beneath the cortex is the dense pericycle, next to which lies the phloem. Thick-walled parenchyma rays, either broad or uniserate, radiate across the xylem, which also contains thick fibers and long vessels. The compound, broad rays are made up of both light- and dark-colored cells. Spring and autumn wood are not well differentiated. Simple, thin-walled parenchyma compose the pith. The young shoot is surrounded by a cuticle that is replaced by periderm (bark) as the stem matures. The epidermis contains numerous stomates that cork over into lenticels and cease to function as the stem ages. Young stems often have longitudinal ridges, whereas more mature stems are nearly circular in cross-section.

Figure 3-2. Two flower buds flank a smaller vegetative bud. Tiny dots along the stem are lenticels. (Source: Gough and Shutak 1978).

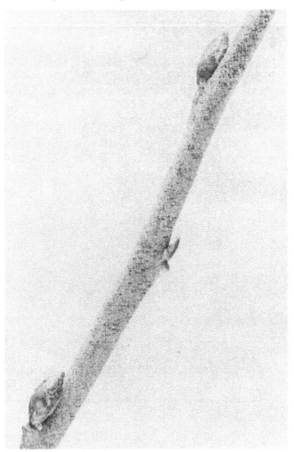

Buds

The dormant vegetative bud is a small, pointed structure approximately 4 mm long and contains a single apex (Gough and Shutak 1978) (Figure 3-2). This vegetative apex is dome-shaped and extends 80-140 μm from the top of the dome to the zone of vacuolated parenchyma near the pith (Figure 3-3). It is about 120 μm in diameter. This is within the range of apical dimensions for most plants. Dormant vegetative apices all have a double-layered tunica 18-20 μm thick. During spring bud swell, axillary bud primordia are initi-

**Figure 3-3. Vegetative apex. Leaf primordium (1), Tunica (2), Vacuolated paren-
chyma of the pith (3). (Source: Gough and Shutak 1978).**

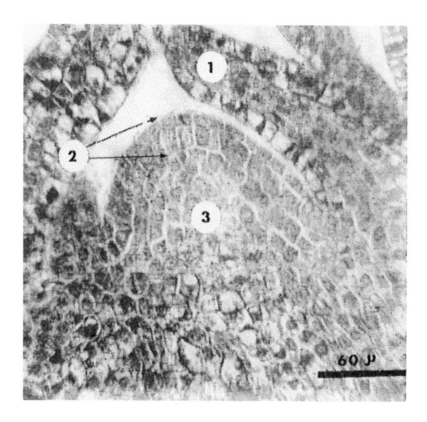

ated along the stem axis, and are thus actually formed more than a
year before they will produce buds. Several bud scales encase and
protect the vegetative apices.

Leaves

Highbush blueberry leaves can reach 75 mm in length and may
have fine hairs on their lower surface, toothed margins, and what
appear to be extrafloral nectaries near their base (Gough et al. 1976).
They are about 2.2 mm thick and contain several structural layers
between the upper and lower epidermis. The upper epidermis is

covered by a cuticle averaging 2.6 µm in thickness, and is composed of simple, transparent cells 24 µm wide. Below this is a double layer of palisade cells. These cells average 64 µm in length by 11 µm in width. A region 150 µm thick containing loosely packed parenchyma cells, called the spongy mesophyll, lies beneath the palisade layer. The palisade cells and spongy mesophyll all contain chloroplasts and are termed, collectively, the mesophyll. The lower epidermis averages 15 µm in thickness and contains about 60 rubiaceous, medium-sized stomates per 100 sq µm. The leaf blade contains both major and minor vascular bundles. The central bundle is composed of one layer of phloem and one of xylem, surrounded by a fiber sheath. Druse crystals containing calcium oxalate are associated with these bundles and are commonly found in highly vacuolated parenchyma toward the upper side of the leaf, between the bundle sheath cells and the primary palisade layer.

Reproductive Anatomy

Flower Buds

Differentiation of flower buds (i.e., the visible change of a vegetative bud into a reproductive bud) begins in mid to late summer (Gough et al. 1978a). The first indication that the process has begun is the flattening of the apical meristem and the appearance of sepal primordia (Figure 3-4). Within an individual bud, the apices differentiate in the sequence in which they are formed, or from bottom to top (acropetal). The differentiating dome-shaped apex flattens, measures about 180 µm in diameter, and is bounded by sepal primordia about 35 µm in width. Petals begin to form within a few weeks. By early fall, all flower parts have formed in centripetal successions (i.e., from the outside edge of the flower to the inside). The carpels occupy a central position surrounded by a ring of filaments, and the ovaries have become highly meristematic by mid autumn. The petals increase in length to enclose completely the perfect portions. Megaspore and microspore mother-cell development is quite active, and cell divisions in that area continue into early winter, then cease until spring, when ovules begin their final stage of development. There is further separation of microsporogenous tissue in the an-

Figure 3-4. Floral differentiation in 'Bluecrop' in late summer. Note sepal primordia (1) and petal primordia (2). (Source: Gough and Shutak 1978).

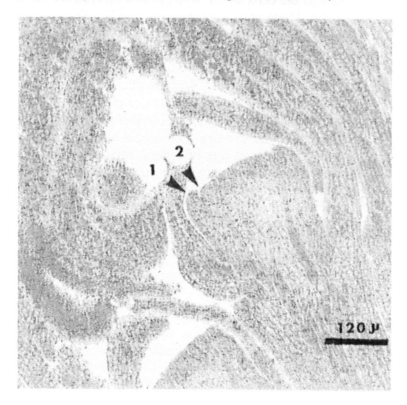

thers, and tapetal disintegration continues until a few weeks prior to bloom, when the individual pollen grains appear (Gough et al. 1978b) (Figure 3-5). Those grains are tricolpate and are shed as tetrahedral tetrads with a diameter of about 30 μm. The sporoderm is undulating and smooth, with a psilate sexine. Just before bloom, the embryo sac is formed and the outer integuments become highly vacuolated, thus indicating the formation of a seed coat. The hollow style of the blueberry is different from the solid style of most angiosperms but is characteristic of the Ericaceae (Figure 3-6).

Differentiation of individual flowers within the bud begins in the lowermost primordia, while the primary apex remains vegetative and continues to cut off additional lateral apices. When all (5-12)

Figure 3-5. Longitudinal section of anthers in 'Bluecrop' prior to bloom. Note disintegrated tapetal cells (1) and pollen grains (2). (Source: Gough et al. 1978b).

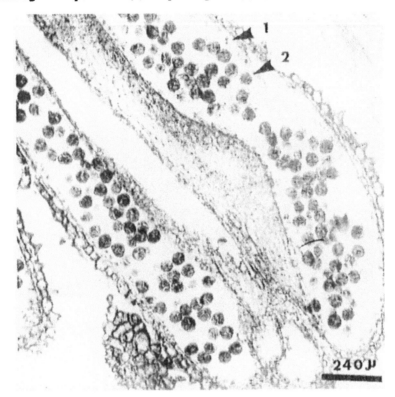

have been formed, the terminal apex aborts, and differentiation of the others occurs in acropetal order until complete. The peduncle elongates, as does each pedicel, resulting in the bud swell that is so apparent during the fall and spring.

Flowers

The early development of the flowers, or florets, within the bud is discussed above. The structure at bloom consists of a white or pink corolla composed of the five fused petals, five fused sepals forming

Figure 3-6. Cross-section of 'Bluecrop' style. Note the darkly stained transmitting tissue lining the inside of the hollow style. (Source: Gough et al. 1978b).

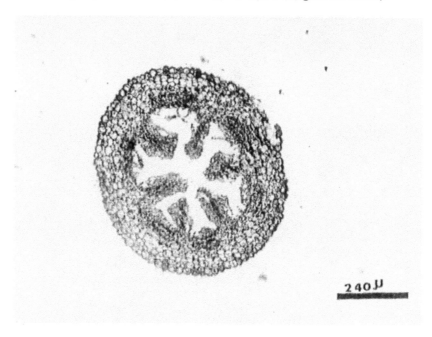

the calyx, eight to ten stamens, and a single style all fused into an inferior ovary (Figure 3-7).

The anatomy of the petals is reminiscent of that of the leaves from which the petals have evolved. An upper and lower epidermis surround a 2-3 layer mesophyll. Both have a cuticle and numerous stomates on their distal (upper) portions. The mesophyll contains large intercellular spaces and a main and numerous smaller vascular bundles containing xylem and phloem.

The sepals are even more leaf-like in that stomates are present on the lower epidermis only, and the thicker mesophyll contains numerous chloroplasts. Otherwise, their anatomy resembles that of the petals.

A hairy, flattened filament supports each elongated anther. Its upper epidermis is covered by a thin cuticle and its cells are elon-

Figure 3-7. Longitudinal section of 'Tifblue' blueberry, × 12. Individual anther (A), × 17; Cross-section of ovary (B), × 12. (Source: McGregor 1976).

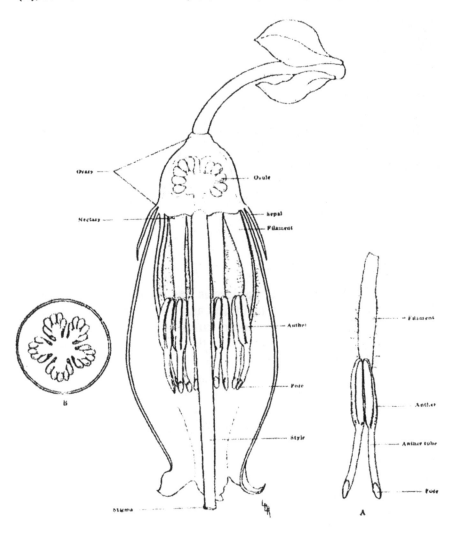

gated into hairs about 500 μm in length. The anther is inserted into the filament at a slight angle, and is adnate with paired awns and apical pores. Because the anther inverts during growth, the pores finally become basal. About two weeks after tetrad formation, each pollen nucleus divides into generative and vegetative portions, and the pollen moves from the anther sacs through two anther tubes and out the terminal pores.

The style, stigma, and ovary make up the pistil. The cylindrical style, like the petals, is made up of an outer and inner epidermis, mesophyll, and conducting tissue. The central lobed, stylar canal is lined with darkly staining, small-celled stigmoid tissue (Gough et al. 1978b) (Figure 3-6). Similar tissue comprises the stigma, a flattened, channeled structure located at the top of the style. The ovary contains ten locules, each connected to placental tissue resting against a central ovarian pillar. The ovary wall is up to 500 μm thick and is made up of chlorophyllous mesophyll between an upper and lower epidermis and contains ten large vascular bundles. The central ovarian pillar contains five bundles, fusing at the style and branching to each placenta, which is comprised of simple parenchyma cells.

Fruit

The single-layered outer epidermis has no stomates and is covered with a cuticle about 5 μm thick and a waxy bloom at maturity (Figure 3-8). The amount of bloom varies with the cultivar and stage of maturity. Pigments are present in the epidermal and hypodermal layers, which are delineated from the rest of the cortex by a ring of vascular bundles. The mesocarp is fairly homogenous parenchyma and contains two more rings of vascular bundles. The carpels contain five large, highly lignified placentae to which up to 65 seeds are attached. The locules, surrounded by a stoney endocarp, extend into the mesocarp. Stone cells are distributed unevenly throughout the mesocarp, but occur with highest frequency from just below the epidermis to a depth of about 1.4 mm (Gough 1983a) (Figure 3-9). They occur rarely in intercarpellary areas and in sepallary tissue, and infrequently in the vascular bundles and in nectariferous tissue (Figure 3-10). Mature stone cells usually appear completely vacuolate and have a thick, heavily pitted, essentially smooth secondary wall composed of several lamellations, each about 1 μm wide. They are composed mostly of lignin and contain little cellulose. The simple wall pits are contiguous with pits in adjacent stone cells, or with pit-fields in adjacent parenchyma walls.

Figure 3-8. Transverse section through a blue fruit of 'Collins' showing darker epidermal (E) and hypodermal (H) layers; the lighter mesocarp area (M); 3 rings of vascular bundles (V); 5 carpels with 10 locules (L); 5 woody placentae (P); and many seeds (S). Extension of locules into the mesocarp usually terminates in a well-defined point (LP). ($\times$ 9.) (Source: Gough 1983a).

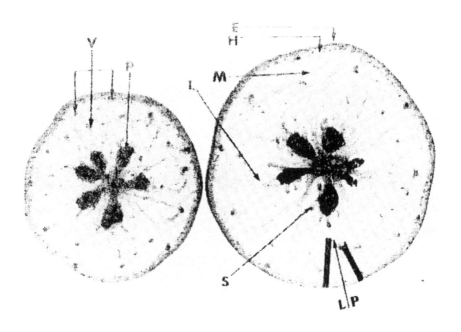

MORPHOLOGY

Vegetative Morphology

Roots

The blueberry root system is primarily composed of fine, fibrous feeder roots less than 2 mm in diameter, along with some thicker anchor and storage roots. The precise distribution of the mass is influenced by plant and soil densities, rainfall, water table, nutrition, etc., and therefore varies somewhat with locale, season, and cultural methods. However, a typical, mature root system has been described

Figure 3-9. Diagrammatic representation of a longitudinal cross-section of blue-berry fruit. Shaded area represents stone cell distribution. Note that few stone cells are present within the nectary ring and none are in the inner portions of the fruit. Heaviest distribution appears to be in the basal lobes surrounding the pedicel scar. (Source: Gough 1983a).

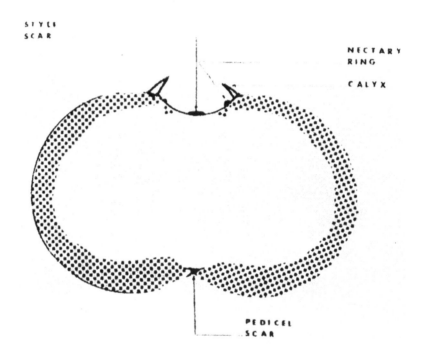

(Gough 1980) (Table 3-1). The root mass is shallow and forms an inverted cone of about 0.32-0.95 cu yd (0.24-0.72 cu m) in volume. Though quite variable, the average dry weight of the total system, excluding the crown, is 3.75 lb (1.7 kg). Roots are mostly parallel to the soil surface. Fine, fibrous roots less than 2 mm in diameter appear in the partially decomposed layer of mulch and become more dense and mingle with larger roots in the upper layers of the soil. Though these larger roots extend to depths of up to 30 in (80 cm), the densest growth occurs in the upper 9 in (23 cm) of soil. The thickest roots are about 11 mm in diameter, but roots greater than 2 mm in diameter compose 30%-60% of the total root dry weight. The

Figure 3-10. Diagrammatic representation of the fruit sector in Figure 3-9. The degree of shading represents the frequency of stone cell distribution. The numbers along the baseline represent distances from the locule point in μm. High distribution frequencies occur within the outer 1400 μm of berry tissue, with highest frequency occurring 460-920 μm below the berry surface. (Source: Gough 1983a).

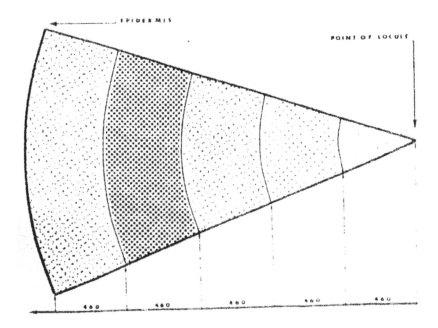

system extends laterally in measurable amounts up to 1.22 yd (1.1 m) from the perimeter of the crown, though traces were noted regularly at distances of up to 2 yd (1.8 m), thus indicating that the systems of adjacent bushes overlap. Overall, about 50% of the root system is located within 12.4 in (31 cm) of the crown, whereas 84% occurs within 24 in (61 cm). The latter distance approximates that between the crown perimeter and the dripline. The blueberry root system, like that of most plants, actually occupies only about 1% of the soil volume of the root zone. Korcak (1992a) found that the root system of highbush blueberry was larger than that of rabbiteye but less efficient in nutrient acquisition.

Table 3-1. Percentage of 'Coville' blueberry roots at various depths and distances from the plant in Bridgehampton fine sandy loam soil with sawdust mulch[z].

| Depth (cm) | Distance from crown (cm) | | | | | | |
	31	61[y]	94	122	153	183	Total
10[x]	--	--	--	--	--	--	--
23	26	15	5	3	T[w]	T	49
36	11	11	5	3	0	0	30
58	11	5	1	1	0	0	18
81	2	3	T	T	0	0	5
Total	50	34	11	7	T	T	

[z]Each figure represents the mean of 6, 13-year-old plants, and is expressed in percent of total dry weight.
[y]Dripline
[x]Depth of mulch
[w]Trace amounts
Source: Gough 1980.

Shoots

The blueberry forms a shrub composed of many shoots arising from buds located on the crown of the plant. The crown is the transition area between the morphologically distinct vascular systems of the root and shoot and, therefore, shows some features intermediate in structure. It develops from the hypocotyl region of the seedling. Shoots develop from the upper side of the crown from buds formed the previous season, or from buds that have remained dormant for long periods. Some vegetative buds form irregularly on older wood. They are called adventitious and have no connection to the apical meristems. Such adventitious buds may develop from wound callus tissue, the cambium, pericycle, or endodermis.

Shoots arise from crown buds that were dormant or adventitious and continue to grow over many years. After their second season, they become quite woody and thickened and are called canes. Some canes arise directly from adventitious buds on the roots and are called root suckers.

During the spring, vegetative bud swell becomes apparent a few weeks before bloom. The swelling is the result of elongation of the stem and axis and initiation of axillary buds and primordial leaves. Since the new internodes are still very short, the leaves will at first appear clustered around the stem (Figure 3-11). The shoot apex continues to produce axillary buds and leaves during stem elongation, which proceeds at different rates on different shoots. Growth of the individual shoots is sympodial and episodic, since it is accompanied by a varying number of apical abortions throughout the season (Gough et al. 1978a; Shutak et al. 1980).

The abortion of the shoot apex and 2 mm of subtending leaf and

Figure 3-11. Early shoot development. The young shoot is still enveloped in leaves. (Source: Gough and Shutak 1978).

stem tissue is commonly termed "black tip." This involves necrosis of the apex and often the bud beneath. The shoot portion between the necrotic area and the next intact axillary bud at first becomes chlorotic. Necrosis proceeds basipetally. Shoots grow in a "flushing" manner (i.e., a shoot will make rapid growth, then stop because of apical abortion) (Figure 3-12). After a short period, an axillary bud near the top of the shoot will break and continue the shoot growth. Each shoot may have a single, or several, growth flushes during the season. Usually, two to five weeks will elapse between flushes, during which time the black tip is sloughed off, leaving a small truncate stem portion above the previously penultimate or, sometimes, antepenultimate bud; the axillary bud that is to continue the shoot growth is then released from its dormancy. The entire

Figure 3-12. Continued vegetative development of a new flush of growth (left). A more advanced stage of development of a new flush (center). Ring of cataphyll scars indicating position of former distal bud prior to its development into a new flush of growth (right). (Source: Gough and Shutak 1978).

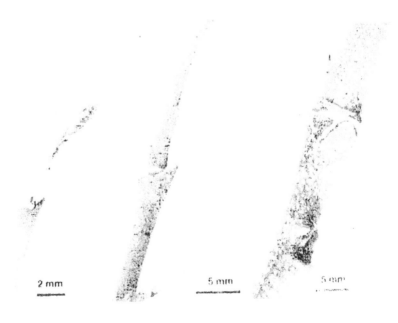

2 mm   5 mm   5 mm

shoot often remains unbranched. However, branched shoots, wherein two or more buds are released after black tip, are not rare. This shoot will continue to grow and is called the second flush. Each flush is terminated by black-tip formation. The number of flushes is cultivar-dependent, but is influenced also by environmental factors (Shutak et al. 1980).

Generally, shoots of cultivars that ripen their fruit early, such as 'Bluetta' and 'Earliblue,' have significantly more flushes than those that ripen their fruit later, such as 'Herbert' and 'Coville.' 'Lateblue' is an exception, behaving more like a mid-season than a late cultivar. Shoot length is strongly influenced by cultural methods. Shoot thickness is also influenced somewhat by the environment, particularly nutrition, though the relative proportions appear to remain the same. In plants receiving standard culture, about 70% of the shoots will be thin, less than 2.5 mm in diameter 1 cm from their base; 25% will be 2.5-5 mm; and 5% will be thick, or greater than 5 mm (Gough and Shutak 1978).

In subsequent years, axillary buds near the base of each shoot or cane usually remain dormant and will not break except under conditions of severe stress, such as heavy pruning. Some never break. Buds near the middle of each cane usually break into shoots of moderate vigor, and those near the top break into vigorous shoots. Buds at the very top differentiate into flower buds. Foliage on shoots near the top of the plant will shade and weaken lower shoots, which then die off or are pruned out over time, often resulting in bushes that are productive only in their canopy and have bare canes with bark-covered buds below.

Leaves

The point of attachment of a leaf to the stem is called a node. The stem area between the nodes is the internode. The angle between the stem and the leaf petiole is termed the leaf axil. Early leaves are formed within the bud along the stem axis. At the time of bud break, up to six leaves may have been formed already and may have buds in their axils (Gough and Shutak 1978). As the shoot grows after bud break, a new leaf is initiated by the shoot apex about every five days in an alternating pattern. The expansion and development of each leaf is controlled by various meristems, and the ultimate leaf shape depends upon the cultivar and environment. Leaves may range in shape from narrow-elliptic to oval. Their upper surface may be dull or glabrous, or rugose or smooth; the margins, entire or serrulate; the mid-vein, smooth or pubescent above or below. The leaves may also have what appear to be extrafloral nectaries. These characters are used in a key for cultivar identification, published by Gough et al. (1976).

Leaf size is affected by the environment and cultivar as well as by shoot thickness. Leaves on thin shoots weigh substantially less than those on thick shoots. Those on medium-thick shoots are intermediate in weight (Shutak et al. 1957).

Leaf number also varies according to shoot thickness. Thin shoots possess about ten leaves, medium-thick shoots about 20 leaves, and thick shoots about 30 leaves (Gough and Shutak 1978).

Reproductive Morphology

Buds

As mentioned above, axillary bud primordia appear microscopically during vegetative bud expansion in the spring, and macroscopically when the young shoots are a few millimeters long. As the shoot axis elongates, the primordia increase in size but remain vegetative until formation of a final black tip and consequent cessation of shoot elongation in late summer (Figure 3-13). At this time, the uppermost bud begins to swell and assumes a terminal position by virtue of displacement of the remaining stem portion immediately beneath the withered apex (Figure 3-14). At about this time, the distal-to-medial portions of the bud scales assume a reddish brown coloration, which continues to intensify as the bud base swells. Such coloration is also apparent on the subjacent bud. As bud swell continues, the outermost cataphylls begin to separate from the bud, exposing underlying cataphylls just beginning to redden (Figure 3-15). The base of the bud continues to swell, displacing the adjacent truncate stem portion. The old leaf buttress has also begun to be displaced. Red coloration develops on the dorsal side of the shoot. The reproductive apex differentiates, as described above, when the bud has expanded to a length of about 5 mm. During bud enlargement, the apex within the bud forms the peduncle of the future inflorescence, initiating axillary meristems as it proceeds. Each of

Figure 3-13. Formation of final black tip. Note necrosis and chlorosis of shoot tip and initial reddening at the tip of bud scales. This is the earliest, most easily identifiable stage of flower bud formation. (Source: Gough and Shutak 1978).

these is linked to the peduncle by its own pedicel. Differentiation of the meristems proceeds acropetally. Peduncle extension finally ends with abortion of the apex, as in typical stem elongation (Gough et al. 1978b) (Figure 3-16). Although the precise time of this abortion is unknown, no vegetative peduncle apices are present in early spring. Flower bud differentiation proceeds down the stem of the current season's wood only. Older wood does not form flower buds.

It has been commonly reported that the blueberry will differentiate flower buds along the upper (distal) portions of the shoot only. In

Figure 3-14. Intermediate stage in the development of a flower bud. Note beginning of reddening near the top of bud scales. Compare with Figures 3-13 (earliest stage) and 3-15. (Source: Gough and Shutak 1978).

general, this is true. Usually, the top half-dozen buds will become flower buds; those in lower axils remain vegetative (Figure 3-17). However, both types are sometimes interspersed along the shoot. For example, on medium-thick shoots having two growth flushes, the uppermost (distal) bud on the first flush is nearly always a flower bud, whereas the next bud up from it, the basal bud on the second flush, is nearly always vegetative (Gough and Shutak 1978). The different types are most often interspersed on thick shoots and least often on thin shoots. Most dispersals occur on the first flush

Figure 3-15. Stage of flower bud development at or immediately prior to floret initiation. Note bud swell and increased reddening of the bud scales and shoot. (Source: Gough and Shutak 1978).

between nodes one and nine (numbered basipetally). When the vegetative and flower buds are interspersed, an odd number of vegetative buds almost always separates two successive flower buds. For example, if a flower bud occurs at node two, then one, three, five, or another odd number of nodes with vegetative buds can be counted before another flower bud occurs (Gough and Shutak 1978).

The number of flower buds on each shoot depends upon the cultivar, but the ratio of flower buds to vegetative buds is related to the shoot thickness (Gough and Shutak 1978) (Figure 3-18). Thick

Figure 3-16. Peduncle of 'Bluecrop' showing black-tip. (Source: Gough et al. 1978b).

shoots, because they are excessively vegetative, have a ratio of 0.35. Thin shoots, because they are weak, have a ratio of 0.55, whereas medium-thick shoots have a ratio of 0.66 and are therefore most productive. Medium-thick shoots also have more flower buds (one) per inch (25 mm) of shoot growth than either thin (0.75) or thick (0.75) shoots. When two flushes of growth are present, the second has more flower buds (one) per inch (25 mm) than the first (0.75).

The flower buds usually occur singly at nodes, though in some cases multiple buds may be present (Shutak 1968) (Figure 3-19).

Figure 3-17. Dormant one-year-old blueberry shoot with flower buds near the top and vegetative buds nearer the base.

Figure 3-18. Three shoots of varying thicknesses. Very vigorous thick (left), productive medium (center), and weak thin (right).

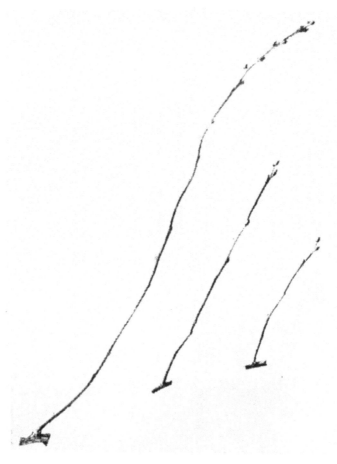

The presence of multiple buds appears to be dependent upon vigor and wood thickness, and independent of cultivar. The minimum requirements seem to be those of medium-thick wood, since the same number might be found on thick and medium wood. Multiple buds are rarely found on thin wood. Multiple flower buds differentiate at about the same time as single buds; the smaller "secondary" buds differentiate simultaneously with, and at the base of, the "primary" buds.

The distal buds contain the greatest number of flowers, and flow-

Figure 3-19. Multiple flower buds on 'Bluecrop.' (Source: Gough and Shutak 1978).

er numbers decrease with increased distance from the tip. For example, the second bud contains an average of nine to ten flowers in 'Collins' and 'Bluecrop,' whereas the third averages eight and the fourth averages seven. It may be that, because flower buds differentiate basipetally along the shoot, the top buds, nearest the apex, have more time to develop before autumn. The secondary buds of a multiple flower bud generally have from one to five flowers. The number of flowers within a bud is cultivar-dependent (Shutak et al. 1980).

Flower

The flowers are arranged singly at the tips of pedicels, which are themselves arranged along a central stem, or peduncle. The peduncle is attached to the shoot and, along with the flowers, forms the typical inflorescence, which in a blueberry is called a cluster or raceme. Pedicels on the same peduncle are not significantly different in length, though length does vary substantially among cultivars. Those whose fruit ripens early generally have shorter pedicels. In structure, the pedicel is quite similar to a stem (Gough, unpublished data).

The flower is generally urn-shaped and inverted during bloom. A glaucous, glabrous, five-lobed calyx is adnate to the inferior ovary. The corolla is usually white, but may be pinkish along the ribs. The degree of pink coloration is due to cultivar, environment, or, sometimes, disease factors. The ovary develops into the berry after the fruit is set.

Fruit

The blueberry fruit is a five-loculed, true berry resulting from the maturation of an inferior ovary. The pericarp is covered with a waxy bloom and is merged with other tissue containing chlorophyll. The boundary between these parts of the ovary can only be distinguished by histological examination. The mature fruit varies from round to oblate in shape; from white, black to light blue, and red in color; and from small to greater than 1 in (2.5 cm) in size. In especially large berries, the segments separating the locules may be visible externally in the form of five ribs. A ring of scar tissue visible immediately inside the calyx and a small dot of the same tissue in the top center of the fruit are the results of abscission of the corolla and style, respectively.

PHYSIOLOGY

Vegetative Physiology

Root Growth

When acceptable conditions are reached in the spring, cells in the root apical meristem begin to divide (Abbott and Gough 1987a).

The new cells are pushed to various areas, including the root cap (where they are eventually sloughed off), and to the zone of elongation located above the meristematic area. Cells in this area elongate, or stretch, and "push" the apex through the soil. Eventually, these cells lose their ability to stretch further and differentiate into various tissues. The faster apical cells divide, the faster the tip is pushed along. As the roots move through the soil, they bend and twist around objects. When growth ceases at the end of the season, the roots brown, or metacutinize, and no longer absorb water.

Older roots undergo a secondary thickening due to the activity of the cambium and phellogen. Roots nearer the surface thicken first, followed by roots deeper down in the soil. Maximum thickening occurs near the soil's surface. Production of xylem within and between roots varies greatly.

Growth Patterns. Under good conditions, a blueberry rootlet may grow up to 1 mm per day (Coville 1910). By comparison, a wheat rootlet may grow 20 times faster. A number of factors influence the rate of root growth. Since they have no true rest period, roots can grow whenever conditions are favorable.

Temperature. Seasonal growth relative to temperature remains nearly constant among cultivars and years (Abbott and Gough 1987a) (Figure 3-20). Little growth occurs when soil temperatures are below 46°F (7°C). When temperatures rise to this threshold in spring, root growth begins. The rate of growth increases until soil temperatures reach about 60°F (16°C) in early summer, then decreases to a low at temperatures above 68°F (20°C). As temperatures again fall, rapid growth resumes until soils again reach 46°F (7°C), after which it stops. Blueberry roots apparently will not grow well in soil temperatures below 46°F (7°C) or above 68°F (20°C). These requirements are similar to those of apples and peaches. In northern areas, root growth undergoes two annual rate peaks—the first at about the time of fruit set and the second during the peak time of flower bud formation.

Water. The mid-summer slump in root growth may also be the result of water deficits. If shoots experience water stress, as when transpiration rates are high and fruit growth is rapid, manufacture of photosynthates and hormones is decreased, and their transport via the phloem to the roots is slowed. A similar situation may occur

Figure 3-20. Elongation of white unsuberized roots in relation to shoot growth, soil temperature, and stage of development of highbush blueberry plants. Vertical lines represent standard errors (SE). SE < 1 are not plotted. Hatched lines indicate beginning and ending of stage; cross-hatched lines indicate peak period of stage. (Source: Abbott and Gough 1987a).

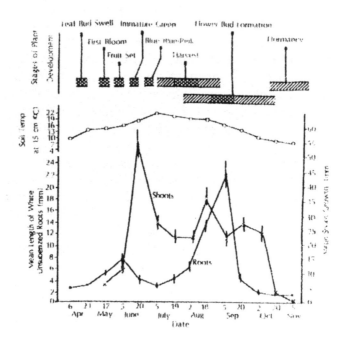

when leaves drop in autumn. Any restriction in the flow of hormones to the roots is more serious than a restriction of carbohydrates. Further, root elongation slows and stops in dry soil, and roots suberize to the tips, thereby reducing their capacity to absorb nutrients and continue growth. Also, a dry soil is resistant to root penetration and, so, slows root growth.

Shoot Growth. It was once widely believed that root and shoot growth were antagonistic. When roots grew rapidly, shoots did not, and vice versa. This may be true for some kinds of plants and under some kinds of deficits, (e.g., when shoots and roots compete for limited substances, such as water); however, Abbott and Gough

(1987a) found no evidence of such antagonism in blueberries under the conditions of their experiment (Figure 3-20). The peak times of root and shoot growth roughly coincided throughout the season. The early peak in shoot growth followed that in root growth by about two weeks, and the second peak preceded that of root growth by about two weeks as well.

Fruit Growth. The presence of ripening fruit may affect the growth of roots (Figure 3-20). The period of slow summer root growth coincides with that of rapid fruit development and ripening. Root growth slows at about the time of fruit set and does not regain its previous rate until harvest. As in other species, the fruit provides a highly competitive sink for carbohydrates and minerals. That is, the plant selectively transports these substances to the fruit rather than to the roots.

Mycorrhizae. The term "mycorrhizae" is taken from the Greek words *myco* ("fungus") and *rhiza* "root"), and it refers to a fungus living in close, symbiotic relationship with a root. In such a relationship, each gives something to the other and no harm comes to either. Many woody plants have such associations. Eynard and Czesnik (1989) reported that blueberry cultivars varied in the amount of mycorrhizae present. They found that 'Berkeley' and 'Herbert' had more than 'Bluecrop' and 'Darrow.' The blueberry root is inhabited by an endotropic form that exists entirely within cortical root cells and does not invade the endodermis or vascular system (Coville 1910). Hyphae protrude into the soil as much as 1 in (2.5 cm) from the root surface and function similarly to root hairs. Although the blueberry plant can grow without this fungus under conditions of high soil fertility, it does much better with it. Evidence suggests that the plant supplies the fungus with carbohydrates and other substances, while the fungus increases the availability of such mineral nutrients as nitrogen, phosphorous, and trace elements to the plant. Roots having this association can absorb more nutrients over a longer period than other roots, and are less subject to attack by root pathogens (Russell 1977). Further, mycorrhizal infections increase a plant's resistance to soil metals by absorbing them within the root cortex and preventing their translocation to shoots (Bradley et al. 1982).

Mycorrhizal associations are retarded by very wet or very dry soils and high fertility, particularly high phosphorous levels. The fungus is present in natural blueberry soils and large-scale inoculation of new, commercial sites may not be feasible, though fruit yield has been stimulated by experimental inoculation of sites in Australia (Powell and Bates 1981).

Nutrient Absorption. Absorption of nutrients from the soil depends upon the root's ability to absorb water, since most nutrients must be dissolved for uptake. Urea and some trace elements are notable exceptions. A molecule of water has two positive hydrogen atoms and one negative oxygen atom. The atoms are arranged so that the positive charges are on one side and the negative charge is on the other. This lopsidedness is called polarity. Because opposite charges attract, like poles of a magnet, the positive side attracts negatively charged particles (anions) and the negative side attracts positively charged particles (cations). Water dissolves various nutrient elements in the soil, separates them into their component positive and negative ions (ionization), and aids in moving them into and through the plant. To obtain water, roots grow into moist soil regions. As the water and nutrients in that region are absorbed, water from surrounding soil replaces them, since water moves from areas of greater to lesser concentration. This same principle allows us to soak up a spill with a sponge. The more water in the soil, the faster it will move into a dry area.

Water and its dissolved ions diffuse through the outer cell walls and enter the root vascular system. The most rapid absorption occurs in the root tip just behind the cap to the area just forward of suberization. This area in the blueberry is also the zone of greatest mycorrhizal association. Some nutrients, however, can be absorbed through older areas of the root. The rapidity of nutrient absorption is a function of the amount of suitable root surface area exposed and of root volume without suberization available. The presence of fungal hyphae greatly increases that surface area. Absorption is slow in older, suberized sections because of water's inability to penetrate the waxy suberin and because of the absence of mycorrhizae. Water diffuses into, and through, the root and into the vascular system (for transport to the rest of the plant). Ions dissociated in water may move in the same manner if their concentration in the root is less than that in the soil. Again, certain ions move from areas of greater to lesser concentration (passive uptake). Water and nutrients appear to move as far as the root cortex by this procedure. Movement through the endodermis into the vascular system is more difficult, however, and may require some energy on the part of the plant (active uptake). This energy is provided by respiration and is used to move ions against a concentration gradient. Therefore, concentration of nutrients in the soil, availability of soil water, and root metabolism all control nutrient absorption.

Plants also have a feedback mechanism that controls absorption of a particular nutrient, based upon plant demand. Simply, if the plant does not need it, it will not be absorbed. Further, absorption of some nutrients may enhance or interfere with the absorption of others. For example, increased phosphate levels enhance nitrate uptake, and excessive levels of cations may decrease uptake of other cations. High levels of potassium ($K+$) reduce the uptake of calcium ($Ca++$) and magnesium ($Mg++$), but may increase the uptake of anions. Furthermore, the presence of large amounts of calcium can reduce the solubility of iron and manganese in the soil and result in "lime-induced chlorosis."

Favorable soil nutrient supply can also cause roots to proliferate. For example, roots will branch rapidly near pockets of fertilizer, particularly nitrogen and phosphorous.

High concentrations of toxic elements such as lead, arsenic, and road salt and improper soil pH can restrict uptake of nutrients. Korcak (1989b) reported that high concentrations of aluminum significantly affected the root tip and restricted growth. Also, unusually high or low soil water content can physically influence the root system as well as alter mineral availability. Phosphorous, potassium, iron, and manganese become particularly unavailable in dry soils, and iron and manganese may become toxic in very wet soils. Soils poorly aerated, either because of compaction or waterlogging, reduce the root's ability to absorb nutrients. Cold soils will also reduce nutrient absorption by the root.

Translocation

The movement of organic or inorganic solutes through the plant is called translocation. As mentioned, movement of water and ions through the root cortex is largely by diffusion, and movement into the vascular system is by active transport. Once nutrients enter the xylem, their translocation up to shoots and leaves depends upon several factors.

Transpiration Stream. There is a continuous thread of water from the soil–via the xylem–to the leaves. This is the transpiration stream. As water vapor is lost through evaporation from leaf stomates, water from within moves into the intercellular spaces in the spongy meso-phyll to replace it. Because of water's ability to stick to itself (cohe-sion), more water molecules are pulled along, like links in a chain. Nutrients move with them and are distributed throughout the plant. The rate of transpiration is influenced by the number of stomates, whether they are opened or closed, and the relative humidity of the atmosphere, temperature, and wind. For example, numerous large holes move more water. A faster wind will move water away from the leaf surface, replacing the moist air with dryer air, into which water vapor can move. Transpiration is very slow in defoliated plants.

Most of the nitrogen utilized by the plant is converted into organ-ic compounds in the roots and translocated to the shoot in the xylem sap, mostly as amino acids. The xylem sap also contains good amounts of growth regulators, enzymes, and other minerals.

Not all ions move at the same rate in the transpiration stream. Calcium, zinc, and iron move more slowly than others.

Desiccation of Leaves. As the growing season progresses, some minerals accumulate in the leaves in the form of salts, which can build to toxic concentrations, particularly when plants are fertilized too heavily. These are usually salts of calcium and magnesium that "burn" the leaves by decreasing the water content of cells.

Remobilization of Nutrients. If not needed where they have been deposited initially, mobile nutrients, such as nitrogen, potassium, and phosphorus, will move out of older tissues and into younger, actively growing areas. This occurs mostly in the phloem. Potas-sium, magnesium, phosphorus, and sulfur are highly phloem-mo-bile, whereas iron, manganese, zinc, molybdenum, and copper are less so. Calcium and boron move very little in the phloem.

Assymetrical Translocation. In many plants, nutrients absorbed from the soil on one side are translocated evenly throughout the plant. This is not the case in the blueberry, however. If water and fertilizer are applied to one side of the plant, then only that side will develop (Figure 3-21). Abbott and Gough (1986) reported that the watered sides of potted plants had substantially greater shoot growth, shoot thickness and length and more leaves and shoots than nonwatered sides. Australian researchers have recently substantiated this response in field plantings (Shelton and Freeman 1989). The root system on the watered sides was also more dense and cohesive than that on the nonwatered side (Figure 3-18). This work

Figure 3-21. Two-year-old potted blueberry plant with split root system and stem straddling plastic partition. Fertilized segment (left), non-fertilized segment (right). (Source: Gough 1984a,b).

was a follow-up to previous work (Gough 1984a,b) in which blue-berry plants were differentially fertilized. The fertilized side responded with significantly better shoot and root growth, bloomed, and set fruit. The nonfertilized side died. Further, red dye injected into the transpiration stream on the fertilized side did not cross over to tissue on the nonfertilized side, thus indicating that there is little cross-transfer of nutrients in blueberry stems.

Shoot Growth

Growth Patterns. Leaf buds begin to increase in size in early spring as a result of combined meristematic activity and internode elongation. Axillary primordia are cut off along the shoot axis and become visible through the microscope at the fourth to sixth leaf back from the apex. Leaf bud opening, or break, can precede that of the flower buds by a few weeks.

The breaking of bud dormancy and the growth of shoots is under the control of several internal and external factors. The external factors include temperature and light (which will be discussed later, in separate chapters). Internal factors include water balance and hormones.

Water Balance. Too much or too little water inhibits bud break and shoot growth. Water is needed to transport hormones and nutrients from root to shoot—and back again—and to regulate meristematic activity. Water balance itself is also responsible for over three-fourths of the variation in shoot growth in woody plants.

Hormones. As growth-promoting hormones increase and growth-inhibiting hormones decrease in spring, buds are released from dormancy. During winter dormancy, levels of gibberellins, a major group of growth-stimulating hormones, and other hormones, such as cytokinens, increase in buds. At the same time, growth inhibitors, such as abscisic acid (ABA), decrease. The interplay between the promoters and the inhibitors controls release from dormancy. Bachelard and Wightman (1974), studying the buds of balsam poplar, divided release into three phases:

Phase I–Gibberellin: Inhibitor ratio increases, along with respiration and mobilization of stored food reserves. Internodes elongate.

Phase II–Hormones translocated to roots. Root activity increases, producing cytokinens and gibberellins, which are translocated to the buds.

Phase III–Cell divisions in the bud are stimulated by cytokinens; cells divide and expand.

The release of bud dormancy in the blueberry may involve a similar process. Subsequent shoot growth is also under the control of hormones. Gibberellins may dominate the interplay of all hormones, including auxins and ABA, thus regulating shoot elongation. Gibberellins are manufactured in newly expanding leaves.

Thickening of the shoots through activity of the cambium is also controlled by auxins, gibberellins, cytokinens, and some growth inhibitors. Because many of these substances are produced by new tissues in leaves and buds, cambium activity generally begins near the top of the shoot and moves downward through older tissue. Therefore, normal growth of the xylem and phloem (normal thickening of the stems) depends upon healthy leaves and buds.

The internal control of shoot growth is extremely complex and not fully understood. Not only does every hormone and related growth substance interact with every other and with changes in the plant itself, but each also interacts with external environmental conditions, although varying degrees of lag time in response do occur.

When considered over the season, blueberry shoots initiate growth a few weeks after the roots (Figure 3-20). They begin slowly and extend rapidly within a month, then slow again, and extend rapidly once more. This recurring sigmoid pattern remains unchanged until shoot growth generally ceases in early fall, and is brought about by the episodic flushing mentioned above. In relation to the fruit, shoots grow rapidly up to fruit set, slow dramatically during ripening, increase growth again after harvest, and finally slow during flower bud formation and the onset of dormancy. Cessation of growth is the result of leaf senescence, changing photoperiod, cooler temperatures, and possibly depletion of nutrient reserves.

Not all shoots begin and end their flushes at the same time. Changes in light, temperature, and moisture and the shoot's position on the bush also influence the growth patterns of each shoot. For example, shaded shoots may be weaker and more "spindly" than those in full sun. Because fruit growth and development and flower bud formation draw heavily upon minerals, carbohydrates, and water in the plant, shoot growth slows considerably when a bush is bearing fruit or is forming flower buds, but is vigorous outside these periods, assuming that all other requirements are adequate.

Epicormic Shoots. Buds that remained dormant from their inception and occupy older areas of the canes (which are usually void of shoots) can be stimulated by a sudden exposure to light to produce epicormic shoots. These are commonly referred to as water sprouts or, incorrectly, suckers. (Suckers arise from roots.) Younger plants often produce more water sprouts, as do plants that have been "opened up" by pruning. Bushes along the perimeter of the plantation may produce more than those at the interior, probably because of better light conditions. Such growth is useful in rejuvenating the plants and in replacing nonproductive wood that has been pruned out. It can, however, result in excessive bushiness if the plant is not thinned properly.

Prolepsis. Some buds that would normally remain dormant until the following spring may break in the fall, producing soft growth that often sustains winter kill. The phenomenon is termed "prolepsis," and the shoot, "proleptic." Formation of these abnormal and usually unproductive shoots is stimulated by heavy autumn rains, late application of nitrogen fertilizer, or improper, late-season irrigation or pruning. Because of proleptics' potential for winter injury, their low productivity, and the resulting "bad form" of their main shoots (because of their tendency to fork), avoid conditions that lead to their formation.

REPRODUCTIVE PHYSIOLOGY

Flower Buds

Initiation

The initiation, or induction, of flower buds is a biochemical signal leading to anatomical alteration (differentiation) of a vegetative apex into a reproductive one (i.e., a chemical command that "tells" the apex to stop producing leaves and begin producing flowers).

Because the exact time of initiation is controlled by weather and cultural practices, it will vary slightly among years and planting sites. Gough et al. (1978a,b) reported that floral differentiation of several cultivars occurred between the end of July and the end of August in the northeastern United States. These dates correspond to between 60 and 90 days after bloom. Actual initiation, therefore, occurred sometime before.

Most fruit and nut species do not initiate flowers in response to photoperiod, although this phenomenon does control other aspects of their growth. Seasonal physiologic age, measured from some event (like date of bloom) or from other factors mentioned above, triggers induction. A blueberry shoot will not form flower buds until it has experienced its last flush of growth and its final apical abortion.

Factors Affecting Initiation

Carbohydrate-to-Nitrogen Ratio

Although the ratio of carbohydrate to nitrogen (C:N) is not primarily responsible for the initiation of flower buds, as once thought, it does influence it. If supplies of nitrogen are excessive in relation to carbohydrates–as would result directly from overfertilization or indirectly from shade, cloudiness, or defoliation because of decreases in photosynthesis–then flower bud formation is decreased. In Florida, defoliation of shoots in August substantially reduced the number of flower buds formed when compared to defoliation of shoots in October (Lyrene 1992). Drought and high temperatures, because they can interfere with nitrogen uptake, can also reduce initiation. Generally, moderate amounts of nitrogen fertilization and strong shoot growth result in the best initiation.

Cool Conditions

Cooler summer temperatures, especially if days are bright, can result in an accumulation of carbohydrates that may increase the number of flower buds formed.

Biennial Bearing

Once common in apples, biennial bearing rarely occurs in blueberries. An exceptionally heavy crop of fruit can drain the plant's supply of carbohydrates and reduce flower bud formation. The following year, only a light crop can be produced, draining fewer nutrients and leaving more available for heavy flower bud production. The cycle is thus perpetuated over the years. This condition has been observed in some cultivars, such as the 'Darrow,' but never as pronounced as in the apple.

Pruning

Severe or improper pruning can result in highly vegetative, lush new growth that will set few flower buds.

Scoring or Ringing

Slicing through the bark only, and completely around the stem, in midsummer will slow vegetative growth and sometimes force flower buds to form.

Bending

Bending branches may encourage flower bud formation, but is impractical on a large scale and results in malformed growth.

Growth Regulators

Some growth regulators, such as SADH, increase flower bud formation in the blueberry (Hapitan et al. 1969; Gough and Shutak 1978).

Development

The impulse for flower bud initiation begins at the shoot apex and moves basipetally (i.e., down the stem). In the northeastern United States, most shoots have completed flower bud formation at their top node by early September. Formation at the fifth node from the top, and at lower nodes, is completed a few weeks after that; the greatest rate of formation occurs in mid-September (Gough and Shutak 1978). As described above, flower buds continue to differentiate and increase in length through the end of October and in width through November. They remain dormant throughout the winter, then begin a rapid increase in size in March. This ends with bloom several weeks later.

FLOWERING

Bloom

Patterns

The opening of flowers and the shedding of the pollen is termed bloom, flowering, or anthesis. When the conditions of chilling have been met and the proper temperature attained, flower buds open. Generally, buds on thin wood open first and those on thick wood last (Hindle et al. 1957). Buds closer to the crown, particularly if bushes have been mulched, open before those near the tops. The flower buds open basipetally along the shoot. That is, those at the top node open first, the second node second, and so on. This sequence is the same as in their formation. The flowers within the bud also open from the tip of the cluster to its base, although they were initiated in the opposite direction. The sequence of flowering holds true for all cultivars. Bud break is the result of rapid elongation of the peduncle and pedicels.

Anthesis

During bloom, the corolla hangs open-end down, and pollen is released during the period of maximum stigma receptivity. Only the stigma at the style tip is receptive. Because the stigma is flared outward, pollen from the same flower is not likely to fall on it. A nectary, or nectar gland, forms a ring at the inside base of the corolla and excretes nectar to attract pollinating insects. Ovary growth up to and during bloom is mostly the result of cell division stimulated in part by substances, probably auxins, formed in the stamens.

Pollen transfer from anther to stigma is called pollination and is the result of bee activity in the blueberry. If pollen from a different kind of plant, such as oak or maple, or from an incompatible cultivar of blueberry, is deposited on the stigma, nothing happens. Rapid changes take place, however, if acceptable pollen is transferred. The pollen is held by a sticky fluid and germinates, sending out a pollen tube that grows along the stylar canals, down into the ovary and ovule, and into the embryo sac, whereupon it ruptures, ejecting two sperm nuclei.

The growth of the pollen tube along the stylar canal is directed by the tube nucleus and is influenced by biochemicals released by the stigmatoid tissue in the canal. The generative nucleus splits into two sperm nuclei, which enter the egg sac. One nucleus fuses with the egg to form a zygote, while the other fuses with two other female polar nuclei to form the endosperm. As the zygote undergoes cell divisions, it becomes the embryo, which is nourished by the endosperm tissue. The outer layers of the ovule begin to harden into a seed coat, and the ovary wall around the developing seed begins to enlarge. Each egg must be fertilized by a separate pollen grain. Given approximately 65 seeds in a mature blueberry, many viable pollen grains must fall on each stigma for acceptable fruit development. Once the eggs have been fertilized, the pedicel begins to turn the flower upward, and the flower parts are shed. The corolla turns brown, then often appears water-soaked before falling. Stamens and styles are shed at about the same time. Only the sepals, fused into a calyx, remain a permanent part of the berry.

FRUIT

Fruit Set

Once fertilization has occurred and the ovary has begun to swell, the fruit is said to have "set." The swelling of the ovary and subsequent fruit set is believed to be due either to (1) pollen's contributing an enzyme that influences auxin or gibberellin production by the stigma or (2) pollen's providing another compound that, along with stigmatic hormones, signals the ovary to swell. Following set, the developing fruit is no longer dependent upon the plant for hormones but instead receives them from the embryo and endosperm in the developing seeds. In fact, the fruit wall will not develop in the area around an unfertilized ovule. This results in "lopsided" fruit. Of course, water and minerals must still be supplied by the parent plant.

Fruit Growth

About two months lapse from anthesis to berry ripening. During this time, berry growth occurs in three distinct stages (Shutak et al. 1957) (Figure 3-22). Stage I is characterized by a rapid increase in

Figure 3-22. Growth curves for fruit of three cultivars. The initial period of rapid growth is followed by a slower, second period just prior to the final rapid swell. (Source: Shutak et al. 1957).

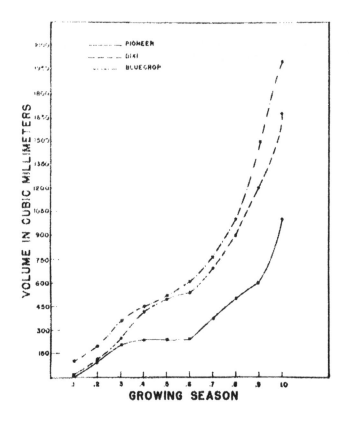

berry volume, attributed mostly to cell division and some enlargement. This stage lasts about a month. During Stage II, the berry size increases little, but the embryos and seeds develop and mature. In Stage III, the berry begins to ripen and undergoes a rapid increase in volume caused by cell enlargement (Galletta 1975). Stage III usually ly lasts 16-26 days and results in the greatest increase in berry size. When volume is plotted over time, growth is marked by a double S-shaped (or double sigmoid) curve. Peaches, figs, and seeded grapes have similar curves.

Large berries experience the shortest Stage II and have about three times as many seeds as small berries (Figure 3-23). In terms of the time allocated to the various stages and to the increases in fruit volume, development of the first third of the total volume takes about 60% of the time; the second third, about 30% of the time; and

Figure 3-23. Representative growth curves for large, medium, and small berries of 'Earliblue.' (Source: Shutak et al. 1980).

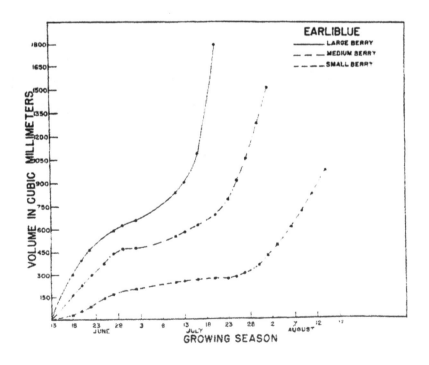

the last third, about 10%. This means that even after a berry turns blue, it may still increase in volume by 25-35%.

Maturation occurs during the period between Stage III of berry growth and ripening. This period is not abruptly defined, however. Physical changes that occur include a softening of the flesh, decrease in chlorophyll and subsequent increase in anthocyanins (i.e., the berry turns from green to blue), and an increase in size (Ballinger et al. 1970). Sugars and other soluble compounds increase in the cortex, acidity decreases, and respiration slowly decreases after an initial sharp upturn, called the climacteric (Windus et al. 1976). There is no correlation between the time a flower opens and the ripening of its fruit, nor is there any pattern of berry ripening within a cluster. Berry size, however, is affected somewhat by the fruit's location on the bush. For example, thicker wood produces larger berries. This may be due to the ability of a greater supply of water and nutrients to get to the fruit because more xylem is available for transport. Further, greater production of photosynthates by the more numerous, thicker leaves on thick wood might also contribute to an increase in fruit size.

Berries on peduncles at the first node are often, but not always, smaller than those on buds further down the shoot (Hindle et al. 1957).

Ripening

Shaded berries usually ripen one to two days later than berries exposed to the sun. This delay appears to be an effect of temperature, since shading can decrease the temperature by about 5°F (3°C) on sunny days (Shutak et al. 1957).

The ripening process has been outlined by Shutak et al. (1980) (Table 3-2). Actual ripening begins when the first color shows, or at about the green-pink (GP) stage, and its progress can be expressed in terms of increases in sugar content. Generally, the sugar content of the fruit will range from about 7% at mature green (MG) to 15% at ripe (R). Earlier ripening fruit tend to be larger and have more sugar than later ripening ones. The increase in sugar is probably due to a combination of sugar accumulation from the plant and to manufacture of sugars within the fruit itself. For example, berries will continue to increase in sugar content after harvest, but will not

Table 3-2. A description of various color stages in berry development.

Color Stage	Description
Immature green (IG)	Hard, completely dark green
Mature green (MG)	Somewhat soft, light green
Green pink (GP)	Some pink showing at calyx
Blue pink (BP)	Mostly blue, some pink at stem end
Blue (B)	Nearly blue, pink ring at stem
Ripe (R)	All blue

Source: Shutak et al. 1980.

attain as high a sugar content as those allowed to ripen fully on the bush. Shutak et al (1980) found that 'Lateblue' fruit harvested at the blue (B) stage contained nearly 12% sugar, which increased to about 13% after five days at room temperature. Fruit that remained on the bush for those five days, however, contained nearly 15% sugar. A berry that has not reached the MG stage will not ripen off the bush.

The rate of water loss (transpiration) varies as berries ripen. MG berries transpire most rapidly and, in general, more mature berries transpire less; however, overripe berries lose water at a very fast rate. Larger berries also transpire at a slower rate than smaller berries, probably because their surface-to-volume ratio is less. Further, fruit of earlier ripening cultivars, such as 'Earliblue' and 'Collins,' transpire faster than those of later ripening cultivars, such as 'Coville.' Cultivars having the highest rates of transpiration also have the highest rates of respiration.

Respiration is the biochemical process of turning stored nutrients into energy. In respiring, the berry utilizes oxygen in the air and releases carbon dioxide, much as humans do. Ovarian respiration is quite high during anthesis, then drops during fruit set. As berries ripen, their respiration, as indicated by CO_2 evolution, increases, reaching a climacteric peak during the initial stages of coloration (GP to blue-pink [BP]) (Windus et al. 1976) (Figure 3-24). Respiration in overripe berries increases once again, probably as a result of tissue breakdown and the onset of senescence. The rate varies by cultivar. As in transpiration, earlier ripening cultivars generally respire faster than later ones.

Figure 3-24. Typical respiration curves of three cultivars of blueberry. (Source: Shutak et al. 1980).

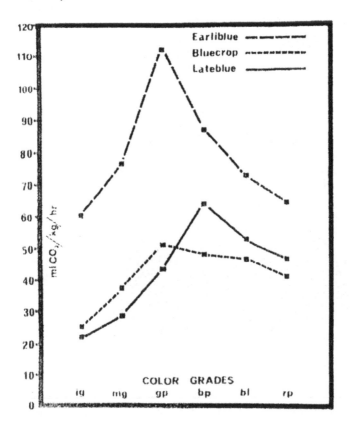

Curves for respiration as a function of ethylene production follow patterns similar to those of CO_2 generation.

Abscission

Final separation of the berry from the bush involves the formation of an abscission zone between the pedicel and the lower, or proximal, end of the fruit (Gough and Litke 1980). At about the GP stage of development, cells in the area between the vascular strands entering the fruit from the pedicel begin to separate (Figure 3-25). Cellu-

Figure 3-25. Microscopic longitudinal section of a blueberry fruit in the green-pink stage of development, showing the berry/pedicel junction and development of the abscission zone. Initial tissue breakdown (B); subepidermal layers (SE). (Source: Gough and Litke 1980).

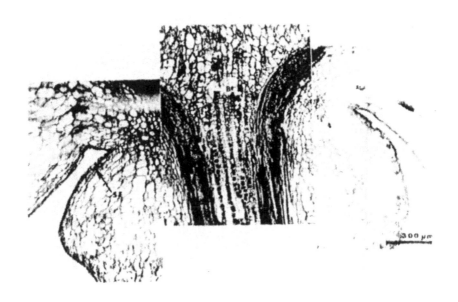

lar breakdown then proceeds in a crescent-shaped area between the berry epidermis and the vascular strands, which themselves do not break down (Figures 3-26 and 3-27). Berry separation is caused by rupture of the cell wall and mechanical tearing of the berry epidermis and vascular strands. This separation leaves a cavity about 5 mm in diameter near its surface, narrowing to 2 mm in diameter at its lowest part, and averaging 2 mm in depth. This is called the scar. The actual size is, in part, genetically controlled and varies among cultivars. Some cultivars are notorious for skin tearing near the scar during berry separation. The degree of ripeness also affects the size of the scar, and the ultimate size influences the fruit's susceptibility to rot. Small, dry scars, as found in 'Bluecrop,' are advantageous. The older cultivar 'Pemberton' has a large, wet scar and so is more susceptible to rot and shriveling.

Figure 3-26. Microscopic longitudinal section of a blueberry fruit early in the ripe stage of development, showing the berry/pedicel junction and development of the abscission zone. Wall fragments (A); tissue breakdown (B). Note tissue breakdown occurs in an approximate crescent from A (on the left side of the figure) through B, and back to A (on the right side), but does not include the dark vascular tissue at the center. (Source: Gough and Litke 1980).

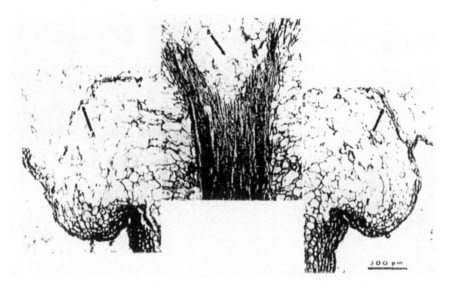

The physiology of abscission is complex and involves the production of a gas hormone, ethylene, within the fruit during maturation. This hormone activates two enzymes, pectin methylesterase and polygalacturonidase, that dissolve the middle lamella (or "glue" that holds cells together), allowing them to separate.

DORMANCY

During the onset of dormancy, visible, active growth ceases. Dormancy can occur at different times in different parts of the plant. In fact, the entire blueberry plant is dormant only during a very short time.

The blueberry plant enters dormancy whenever growth is impossible, either because of drought or, more commonly, cold. Seeds may undergo several types of dormancy. The shoots of blueberry

Figure 3-27. Microscopic longitudinal section of a blueberry fruit in the full ripe stage of development, showing nearly complete breakdown of tissue in the abscission zone. The berry remains attached to the stem only by the vascular tissue. Note fully developed stone cells (BA). (Source: Gough and Litke 1980).

plants pass through different types of dormancy—namely, correlated inhibition, quiescence, and rest. The names for the types of dormancy change constantly, but this scheme will suffice for discussion here. More than one of these may occur within the plant at the same time.

Correlated Inhibition

This is a physiological dormancy maintained completely by agents within the plant but not within the dormant organ itself. For example, axillary buds do not usually grow the first year. Their fully formed primordia remain dormant until the second season, then elongate into shoots. The main shoot apex releases a growth-inhibiting hormone that is translocated down the stem to the axillary buds. As long as the main apex functions, axillary buds will not grow. During apical abortion, however, the main apex is destroyed. Because the uppermost axillary bud is the first to be removed from its influence, it begins to grow, assumes dominance, and releases the same hormone to inhibit growth of the other buds. Sometimes, the upper two buds are released, resulting in a forked shoot. This type of dormancy is called "correlated inhibition," because the axillary bud inhibition is correlated with apical activity. The entire plant may be growing rapidly, with only some parts of individual shoots affected.

Quiescence

This is a type of dormancy imposed upon the plant by external conditions. It is fleeting, and once the environmental stimulus inhibiting growth is removed, the plant will grow again. Bushes suffering from drought will cease growing. Once soil moisture levels return to normal, growth resumes. Temperatures excessively high or low, beyond the cardinal temperatures, will stop growth and force the plant into quiescence. When temperatures again occur within the proper range, the plant resumes its activity.

Rest

This is another type of physiological dormancy. Like quiescence, the stimulus for rest originates in the environment external to the plant, but, like correlated inhibition, it is maintained by factors within the plant. Once these factors are set in motion, the plant becomes dormant and will not resume growth until certain internal conditions have been met, even though external conditions may become conducive to growth once again.

A resting blueberry shoot will not grow when brought into a greenhouse in late fall, but it will grow if brought in in late winter. Conditions in the greenhouse are the same in both cases; only the conditions inside the buds are different. The requirements of rest are of great importance to blueberry growers and should be well understood.

There are three phases to rest. During the onset of rest in the fall, shoot growth slows and stops, followed by cessation of flower bud activity. Onset is gradual, taking several weeks. During this time, the bush becomes less and less responsive to environmental stimuli. For example, heavy irrigation or applications of nitrogen in early fall can result in lush shoot growth. Similar applications in mid fall result in only moderate growth, whereas in late fall, no growth will be observed. The plant's systems are gradually shut down for the next phase, which is known as deep rest. During this phase, the above-ground part of the plant is completely unresponsive to environmental stimuli. Shoots brought into the greenhouse at this time will not grow, no matter how conducive the conditions. Once into this phase, the shoot must be exposed to a certain number of hours of temperatures below 45°F (7°C) to resume growth. The chilling requirement varies somewhat among cultivars, but ranges from a low of less than 400 hours for southern, or low-chill, highbush adapted to central Florida conditions, to nearly 1100 hours for best growth of some northern cultivars.

In general, the more chilling a cultivar receives, the stronger the spring growth will be. Insufficient chilling will lead to poor bud break and fruit development. Because flower buds require somewhat less chilling than leaf buds, they often break earlier in the spring. Buds on very vigorous shoots require more chilling than those on weaker shoots. Greater cold will not necessarily result in accumulation of more chilling hours, however. Temperatures of 35°-55°F (1.5°-12.4°C) are effective in satisfying the chilling requirement, though temperatures of 37°-49°F (2.5°-9.1°C) are most effective. Temperatures below 35°F (1.6°C) or above 55°F (12.4°C) have little effect, and temperatures higher than 61°F (16°C) actually have a negative effect (Norvell and Moore 1982).

The hours required for satisfactory chilling are usually accumulated before the end of winter. Indeed, cultivars should be selected on this basis and planted in areas where this will occur. Planting northern cultivars in southern areas is not appropriate.

Once the chilling requirement has been satisfied, the bush can again grow, provided external conditions are conducive. If they are not, the plant may be released from rest but will remain in quiescence until conditions are right.

The exact trigger for the onset of rest is unknown, but it probably involves a combination of factors, such as shorter day lengths and lower temperatures. These factors, in turn, trigger internal changes in hormones and general metabolism. Growth inhibitors, such as abscisic acid, increase; growth promoters, such as the gibberellins and auxins, along with respiration, decrease.

Roots do not undergo a true rest, but remain quiescent during winter months. This quiescence is induced by both low soil temperatures and high water tensions, since water in the frozen form is not available for absorption. Since winter root-zone temperatures are often higher than ambient air temperatures, however, roots may continue to grow into midwinter, even though shoot growth ceased several months before (Abbott and Gough 1987a).

Under normal conditions, blueberry plants must complete their rest requirement in order to grow properly; however, rest is not absolutely needed for blueberry plant survival (Gough, unpublished data). Although it is true that the blueberry must satisfy its chilling requirement to break rest, that requirement is circumvented if the plant never enters rest. In greenhouse experiments, the author subjected potted plants to continuous early-summer photoperiods and temperatures and found that the plants continued to grow and produce over a two-year period, without entering dormancy. Shoot growth, flower buds, flowers, and fruit were present concurrently on the same plant. Although the peduncles more closely resembled shoots, the pedicels were greatly elongated (bearing the single berry at their tip), fruit did set and develop acceptable flavor.

During the onset of rest, the blueberry plant becomes acclimated slowly to the coming cold period. Because it is adapted to northern regions, the plant will tolerate temperatures well below freezing, provided it has first acclimated, or hardened. During this gradual process, changes occur in cellular proteins, fats, carbohydrates, organic acids, amino acids, nucleic acids, and internal water levels that allow the plant to survive low temperatures. These changes will not occur properly in weak plants.

Hardening

Hardening is achieved in two stages. The first begins when growth has stopped in early autumn, before the plant enters rest. It appears to be triggered by short day lengths, which signal the leaves to manufacture a hardiness promoter. This promoter, which is probably a hormone, is translocated throughout the stem when temperatures are cool. Cold temperatures restrict this process. The second stage requires a frost or, more properly, temperatures below freezing. With the first frost, certain enzyme systems are probably activated that further harden the tissues. The stimulus for this second stage does not move throughout the plant. Temperatures above freezing quickly reduce the degree of tissue hardiness and render the plant susceptible to subsequent cold damage. This "dehardening" process, which occurs during midwinter thaws, is not conducive to the overall health of the plantation.

Different parts of the plant vary in the degree of hardiness they can obtain. For example, well-acclimated canes of 'Earliblue' can withstand an extreme of –40°F (–40°C). Because of fluctuating winter temperature, however, some plant damage usually occurs when temperatures fall to around –20°F (–29°C).

Costante and Boyce (1968) found that cultivars harden rapidly during early fall, then more slowly. They reach their maximum hardiness in midwinter, then deharden gradually to bloom. The greatest degree of hardiness corresponds to the period of deepest rest. Dehardening begins when rest is completed.

Flower buds can also withstand temperatures of –40°F (–40°C). This ability is directly related to the presence or absence of water in the tissues. Under extremely cold conditions and low atmospheric humidity, water is drawn out of flower tissues. Since there is little water left to freeze, damage from ice crystal formation is lessened. The difference in the ability of flowers of different cultivars to lose water results in their differences to sustain potential injury.

Considerable injury can also occur to the vascular tissue connecting each bud to the stem. When this is damaged, the flow of water, nutrients, and hormones to the bud developing in the spring is restricted, resulting in death of that bud soon after spring growth begins. Shoots may even leaf out and set fruit, only to shrivel and die soon after because of the winter injury to the vascular tissue.

The location of the flower bud on the stem and its stage of development also influence the bud's ability to escape cold damage. For example, buds at the tips of shoots do not harden well and are quite often killed each year. Flower buds farther down the branch are less developed and more hardy (Hancock et al. 1987). Also, flowers at the top of the clusters are damaged more easily than those near the base. As the season advances toward bloom, buds deharden and can be killed by successively higher temperatures.

Roots do not enter a true rest period and do not harden as much as the ariel portions of the plant. They are damaged at higher temperatures, ranging from 5° to 20°F (–15° to –7°C). These temperatures are those of the soil in the root zone. Small roots are less hardy than larger roots. Roots left partially uncovered throughout the year may tolerate considerably lower temperatures. Good soil management will protect the roots from serious damage.

Factors that promote hardening are those that result in early cessation of growth and accumulation of food reserves in the canes and storage roots. These conditions are outlined in Table 3-3.

Although a plant's ability to withstand cold depends upon numerous factors, cultivars can, for practical purposes, be ranked as shown in Table 3-4.

Cold Damage

Because different parts of the plant react differently to the hardening process, forms of injury vary according to when killing temperatures occur.

Fall Cold Damage

A rapid drop in temperature during the early fall, before plants have entered deep rest, often results in the death of some shoot tips. These are the tips that either have not quite stopped growing or have not yet "hardened-off." Any condition that stimulates late-season

Table 3-3. Factors which influence hardening in highbush blueberry.

Promote Hardening	Delay Hardening
Low nitrogen	High nitrogen
Drier soil	Wetter soil
Large leaf area	Weak shoots, small leaf area
Light to moderate crop	Heavy crop
Good soil drainage	Poor soil drainage
Full sunlight	Shade
Good leaf retention	Early defoliation
Cover crops, mulch, sod	Clean cultivation

Table 3-4. A partial list of cultivars in approximate grouping by their hardiness.[z]

Hardy	Medium Hardy	Least Hardy
Patriot	Jersey	Berkeley
Northland	Burlington	Coville
Meader	Rubel	Pemberton
Bluecrop	Earliblue	Dixi
Blueray	Rancocas	Stanley
Elliott	Weymouth	Concord
Herbert	Atlantic	Murphy
Bluetta	Collins	Wolcott
Northcountry		Morrow
Northblue		Croatan
Northsky		South. High.
Friendship		

[z]Much of this ranking is adapted from publications of the United States Department of Agriculture. Other researchers may rank cultivars in a slightly different manner.

shoot growth–such as continued irrigation, late fertilizer application, heavy application of slow-release fertilizers and/or manures, or poorly timed pruning–increases the chances of this type of winter damage.

In areas with fairly cold temperatures during early fall, followed by warmer temperatures, the top few flower buds on plants of some cultivars will bloom and may even set fruit. This habit appears positively correlated to the amount of lowbush blueberry in the cultivar's parentage (Fear et al. 1985). It is not uncommon to have green fruit on some plants in some areas on Thanksgiving. The flowers and fruit are killed by subsequent cold temperatures, and may represent a loss to the grower of from 2 to 4 pt (1-2 l) per bush. In New England, 'Bluetta' may often have more than 2-3 flower buds per shoot open in mid-November; 'Earliblue' and 'Patriot,' 1-2; and 'Collins' and 'Darrow,' a few on each bush (Gough, unpublished data). Since the grower can do little to correct this problem, overplanting these cultivars is not recommended in areas where warm falls are common.

Winter Cold Damage

Damage during winter is commonly the result of prolonged extreme cold (below −25°F (−32°C) or of fluctuating temperatures. The latter case is particularly true after January, because the plant has satisfied its rest period and is more responsive to warm temperatures. Therefore, plants in locations that experience a "midwinter thaw" followed by a rapid drop in temperature may sustain substantial damage to the upper portion of shoots. Growers in areas that frequently experience extreme cold with little snow cover, or prolonged midwinter thaws, should strongly reconsider the advisability of planting the highbush blueberry.

Spring Cold Damage

A sudden, extreme drop in temperature when plants are emerging from quiescence may result in substantial bud damage. The major concern during this period, however, is the cold temperatures during bloom. As the blueberry flower bud begins to swell and flowers begin to emerge, the tissues become increasingly susceptible to cold damage. Depending upon the cultivar and the preceding temperatures, flowers in full bloom may be damaged at temperatures of 23°-28°F (−5° to −2°C). If the temperature drops rapidly from relatively mild temperatures, injury may occur at even higher temperatures. If, however, cool weather occurs throughout the bloom period, flowers may not be damaged until the lower limit is reached. Since not all buds are in the same stage of development at the same time, bud position on the shoot is significantly correlated with susceptibility to frost damage. Hancock et al. (1987) reported that buds one and five often vary by a full developmental stage.

The ariel portions of the plant may be affected rapidly by changing conditions; however, root damage is usually the result of prolonged extreme cold over an entire season. This situation is made worse when the soil surface is left bare. Slight damage will often go unnoticed, but plants with severely damaged root systems may leaf out in the spring and even set some fruit, then suddenly wilt and die.

Chapter 4

Climatic Requirements

TEMPERATURE

Growing Season

In a general sense, growing season refers to the time of year between the last spring frost and the first fall frost. This description is somewhat misleading, however, since a blueberry plant will not grow when the temperature is below 38°F (3°C), even though there is no frost. The latest ripening cultivars require a growing season of at least 160 days (Kender and Brightwell 1966). Some early and midseason cultivars may ripen their fruit in only 120-140 days, but areas with such short seasons are marginal for good, commercial production.

The importance of temperature is primary in any discussion of growing season, since the number of days available for growth is not as important as how hot those days are. According to Hopkins' bioclimatic law, the growing season begins four days later, and the temperature is reduced by 1°F, for each degree of north or south latitude away from the equator or each 400 ft (130 m) rise in elevation. Daylength increases with increasing northern latitude during summer, however, thereby somewhat offsetting potentially colder temperatures. Because large bodies of water hold more heat and release it more slowly than air, conditions along the coasts and around large lakes are more moderate than those near the interior of continents.

Temperature affects nearly all the biological processes of the plant and, within the range mentioned above (46°–68°F) (8°–20°C), the higher the temperature, the faster the plant will grow and the fruit ripen. Given optimum supplies of water, nutrients, and light, a plant's growth rate will approximately double for every 18°F (10°C) rise in temperature.

Climatic requirements also involve not only the duration of appropriate temperatures for growth but also the temperature extremes. Blueberry plants have been hardened to withstand –40°F (–40°C) in deep rest (Quamme et al. 1972); however, since winter temperatures often rise above 28°F (–2.2°C), resulting in some potential for dehardening, blueberries should not be planted where winter temperatures regularly fall below –20°F (–29°C) (Kender and Brightwell 1966). An exception to this might be the use of newer half-high cultivars, such as 'Northland' and 'Northblue.' Because of their low stature and flexible canes, which bend under heavy snow, they are covered for most of the winter and, thus, are insulated from extreme cold. Wildung and Sargent (1989a) reported that 'Northblue' bore a normal crop with a minimum of 12 in (30 cm) of snow cover, which protected the plant from winter temperatures of –34°F (–37°C). Less than 6 in (15 cm) of snow resulted in nearly complete dieback and 100% fruit loss. Plants with less than 40% dieback compensated for loss of fruiting wood by producing larger berries. Total yield was unaffected. Where snow cover is unusually light, row covers can insulate the plants and have the potential to double yields over those on unprotected plants (Wildung and Sargent 1989b). White polyethylene, spunbonded material, and 0.12 in (0.3 cm) mesh netting all gave adequate protection to –34°F (–37°C). However, plants under the plastic and spunbonded material bloomed eight to ten days earlier, increasing the potential for frost damage. Areas normally experiencing extremely cold temperatures and little snow are not suitable for blueberry production.

The blueberry can survive temperatures up to 120°F (50°C) for only short periods. Leaf temperatures above about 86°F (30°C) can cause internal water deficits, sunburn, chlorosis, death of the phloem and cambium, uneven ripening, and poor coloration (Hartmann et al. 1988). Growth will stop at these temperatures and will slow considerably at leaf temperatures above 68°F (20°C). Leaves in full sun may be up to 27°F (15°C) warmer than the air, and some fruit, particularly those with small surface-to-volume ratios, may have a temperature up to 54°F (28°C) above the air temperature. The plants die under very high temperatures because the rate of water loss exceeds the rate of water uptake (Kender and Brightwell 1966). Hancock et al. (1992) reported that, while photosynthesis was reduced at 86°F (30°C) there was a difference among cultivars in their ability to tolerate heat. 'Jersey,' 'Elliott,' and 'Rubel' were significantly more tolerant than 'Spartan,' 'Bluejay,' and 'Patriot.'

Suitable air temperatures are, of course, necessary to satisfy the plant's chilling requirement and can influence soil temperatures, which in turn influence the growth of the plant. For example, with cool soil temperatures below 68°F (20°C), some cultivars form short, spreading bushes, whereas higher soil temperatures result in taller, lankier plants (Bailey and Jones 1941). Furthermore, flower bud formation is greater at temperatures around 75°F (24°C) than at 60°F (16°C). Warmer temperatures are also conducive to faster pollen germination and tube growth, and to better fruit set and faster ripening (Knight and Scott 1964; Shutak et al. 1956). Temperatures approaching 90°F (32°C) during ripening, however, can result in smaller, less flavorsome, bland fruit with bloom more subject to rubbing (Mainland 1989). Cold springs not only delay bloom, but also slow root growth and nutrient absorption. This slowing may cause some temporary leaf discoloration. New leaves on emerging shoots of most cultivars will be slightly reddened, whereas those on 'Earliblue' will turn orange-yellow. This discoloration disappears soon after temperatures rise.

Autumn frosts are of little concern, but spring frost can substantially influence production. The degree of injury is influenced by the stage of bud or flower development; the amount of leaf cover over the blossoms; the temperatures preceding the freeze; the freeze severity and duration; the wind speed; cloud cover; and surface moisture. In general, flowers in full bloom are killed at 32°F (0°C). Buds that are fully expanded and show individual, unopened flowers are killed at 28°F (−2°C), and those that are somewhat swollen but not yet opened are killed at 21°F (−6°C). The degree of damage can be estimated by recording the number of blackened pistils and ovaries several hours after the temperature has warmed following a frost. Where injury is suspected to have occurred during winter months, flower buds can be warmed at room temperature for several days, then cut open and examined for browned tissue.

Leaves covering blossoms can somewhat protect them from frost damage by reducing the amount of heat lost through radiation from the flower's surface. The critical frost period usually occurs before substantial leaf cover has developed, however. If the temperatures have been generally cool preceding a frost, blossoms may be slightly more tolerant of the low temperatures. Furthermore, the cooler preceding temperatures will have delayed development of the bush so that buds and blossoms will be in a less developed, less susceptible stage. A rapid drop in temperature, followed by prolonged, severe cold, is more destructive than a short dip to temperatures just below freezing. A gentle wind may reduce frosts caused by temperature inversions by mixing the air; but in cases of mass-flow freezes, wind may increase the damage by dissipating any radiant heat remaining near the plant. Cloud cover keeps radiant heat from escaping into space, thus keeping the air surrounding a plantation warmer. Because of its high latent heat, water traps more heat than air and releases it more slowly. Wet soil releases up to 2.5 times more heat into the surrounding air than does dry soil.

Temperatures below 32°F (0°C) may result in moisture freezing on plant surfaces ("hoar frost"), or may not ("black frost"). Freezing temperatures and the presence or absence of frost are both critical to production. Frosts and freezes can occur in several ways:

Radiational Cooling

The earth accumulates heat from the sun during the day and radiates it back into the atmosphere at night. Cloud cover reflects part of the escaping heat back toward the earth, in the same way a blanket conserves body heat. On clear nights, heat is radiated into space and the atmosphere cools rapidly, setting conditions for a radiational frost. Calm nights, with little mixing of warm and cold air, are particularly dangerous.

Air-Mass Freeze

Frost may also occur when a mass of air with subfreezing temperatures flows over an area.

Temperature Inversion

Temperature inversion occurs when a heavier, cold air mass is sandwiched between a lighter, warm air mass and the soil's surface.

Cloudy, windless nights are particularly conducive to this type of freeze. A low-ceiling inversion occurs when the layer of warm air extends nearly to plant-top level. In a high-ceiling inversion, the warm air layer is considerably higher. Since any heat added by management practices will rise to air with a similar temperature, it is much less costly to protect the crop during periods of low-ceiling inversions.

Potential frost damage may be lessened either by the addition of heat or by reduction of its rate of loss. Reduction of heat loss can be accomplished by covering the plants, though this is impractical in most cases. Addition of heat, the method most employed by growers, requires energy input and can be costly. Plantation heaters, such as infrared devices or smudge pots (banned in many areas), require tremendous amounts of energy or are quite environmentally unsound. Wind machines, which are essentially propellers on poles, circulate the air over the plantation, mixing the higher, warmer air with the lower, colder air. These are particularly useful during temperature inversions, when air 50 ft (15 m) above the soil surface may be up to 10°F (6°C) warmer than that near the bushes. They are of little use in reducing the incidence of frosts induced by a cold air mass.

Soil management systems have an effect on frost damage. For example, bare, moist, firm ground holds a lot of the heat that is radiated to the plant canopy at night. Moist soil under a low cover crop or stubble, dry, firm ground, and freshly disked soil stays about 0.5-3°F (about 1°C) cooler than bare moist soil. A high cover crop keeps the soil up to 4°F (2°C) cooler, and a very high weed cover may keep it up to 8°F (2.5°C) cooler than a bare, moist surface (Galletta and Himelrick 1990). In all these cases, the soil is increasingly unable to give up heat to the plants during a potential frost. On the other hand, the plants are also in a less susceptible stage of development because of the cooler temperatures.

Overhead sprinklers are perhaps the most widely used protection against frosts. They are efficient and can also be used during the growing season for irrigation. Water is a reservoir of heat. When an ounce (28 g) of it freezes, it releases 2,240 calories (heat of fusion). Therefore, as water freezes continuously, an uninterrupted supply of heat is furnished to the plant. This is not a great amount of heat, but it may keep flower tissues above the critical temperature, protecting blossoms to 20°F (–7°C). Because it is the freezing of water that supplies heat, and not the resulting coat of ice, the sprinklers should remain running until temperatures rise and the ice melts. The resulting ice build-up may itself cause some cane breakage and mechanical damage to the flowers, however. Table 4-1 gives the sprinkling rate necessary to protect flowers from cold. Be sure to activate the sprinklers before temperatures become critical. Because wind and very low humidity increase evaporation, which cools tissue, sprinklers offer best protection from radiation frosts (Westwood 1988). Use of sprinklers during bloom in southern growing regions can increase the incidence of *Botrytis* (P. Lyrene, personal communication).

LIGHT

Light Quality

No higher plant can grow for very long without light. Light refers to the small, visible portion of the electromagnetic spectrum of

Table 4-1. Sprinkling rate necessary for frost protection during bloom.

Temperature of dry leaf[z]		Wind Speed (mph)					
F	C	0-1	2-4	5-8	10-14	18-22	30
27	–3	0.10[y]	0.10	0.10	0.10	0.20	0.30
26	–3	0.10	0.10	0.14	0.20	0.40	0.60
24	–4	0.10	0.16	0.30	0.40	0.80	1.60
22	–6	0.12	0.24	0.50	0.60	1.20	1.80
20	–7	0.16	0.30	0.60	0.80	1.60	2.40
18	–8	0.20	0.40	0.70	1.00	2.00	3.00
15	–9	0.26	0.50	0.90	1.30	2.60	4.00
11	–12	0.34	0.70	1.20	1.70	3.40	5.00

[z]Estimated temperature range: from 1 F below the air temperature, with light wind, to 3-4 F (1.5 C), with no wind.
[y]inches per hour
Source: Gerber 1970.

energy we receive from the sun. The wavelength of any particular portion of this spectrum determines the nature of the radiation. Cosmic rays, X rays, and ultraviolet rays, none of which we can see, have very short wavelengths. These, and radiation with very long wavelengths (again, which we cannot see), are not useful to plants. Wavelengths between 390 and 760 nm power the plant world. The light we perceive from the sun is really a mixture of all visible wavelengths, yet it appears white; however, certain component colors in light regulate certain growth processes.

Photosynthesis is an important plant process. In it, carbon dioxide and water are converted into sugar in the presence of the green pigment (chlorophyll) and light in the blue and red-orange wavelengths. The sugar is then used in general growth and metabolism, such as root growth, flower bud formation, and fruit production.

The bending of shoots or whole plants toward the sun and the stretching, or etiolation, of shoots under low-light conditions are both governed by blue light. Blue light also opens the stomates and regulates the acidity of the guard cells, which in turn regulate the conversion of starch to sugar. The absence of light alters the acidity of the guard cells once again, sugar is converted back into starch, and the stomates close, thus regulating photosynthesis.

Light Intensity

Bright moonlight delivers about 0.02 foot-candle (fc) (0.22 lux) of light. To read this book comfortably, your eyes require about 20 fc (215 lux). At high noon on a bright summer day, about 10,000 fc (107,640 lux) hit the tops of blueberry plants in full sun. Photochemical processes capture about 1% of this energy. The leaves of the inner canopy of a blueberry bush, however, receive only about 600 fc (6,458 lux), and are therefore less productive (Teramura et al. 1979). Shoots of the inner canopy may even weaken and die, thus resulting in long canes that have useful leaf and bearing surfaces only on their upper end.

The blueberry bush only needs about 1,000 fc (10,764 lux) to saturate its photosynthetic mechanism, but a reduction in light intensity to 650 fc (6,997 lux) will significantly reduce photosynthesis.

In some blueberries, the process of flower bud formation requires more than 2500 fc (26,910 lux) (Hall and Ludwig 1961). At intensities under 200-300 fc (2,153-3,229 lux), flower bud formation and vegetative growth are substantially reduced, and during the following season berries will develop abnormally and drop prematurely (Shutak 1966). Planting blueberries in full sun will help both photosynthesis and flower bud formation to proceed at the fastest rates.

The distribution of light throughout the day and the season controls the photoperiodic response in the plant. The daylength, or photoperiod, is "perceived" by dormant buds in the presence of the growth-controlling hormone phytochrome. This pigment responds to daylength by "ordering" metabolic changes to promote certain types of growth. For example, the blueberry plant will remain vegetative and form no flower buds under daylengths of 16 hours. (Hall et al. 1963). As the daylength shortens, vegetative buds will differentiate into flower buds; maximum formation occurs during daylengths of under eight hours. In lowbush blueberry, the short days must continue for at least six weeks for normal flower bud formation (Hall and Ludwig 1961). In some cases, cool temperatures can substitute in part for short days.

The grower must remember that any discussion of light should always bring to mind heat, since the brighter days are often warmer and since light and heat are both part of the electromagnetic spectrum.

WATER

Blueberries are somewhat drought-tolerant; however, growth of rabbiteye blueberry is reduced under conditions of only moderate water stress (Davies and Johnson 1982). Even symptoms of incipient wilt are too severe for the best production. The plant requires a constant supply of water from spring bud break to autumn leaf fall. Under northern conditions, about 1 in (25 mm) of water per week will satisfy the minimum requirement, though 1.5 in (40 mm) are needed for the best production from fruit set to harvest. Much of this is supplied by normal rainfall, although during droughts, supplemental irrigation is necessary. The amount of water supplied depends upon the water-holding capacity of the soil and upon prevalence of disease. In Florida, where *Phytophthora* is a problem, growers are urged to apply about 1.5 in (40 mm) of water over a week's time and then no more for three weeks, to allow some soil drying (P. Lyrene, personal communication).

Growers often employ the "feel test" to determine moisture needs. A sample of soil from the root zone is squeezed in the fist. If the soil forms a weak ball that breaks easily, enough moisture is available for good growth. If the soil will not form a ball, irrigation is indicated. If the soil ball is not easily broken, or if water can be squeezed from it, it is too wet. Considerable experience and knowledge of soil types are necessary to use this test properly. Remember to sample soil from the root zone, not the surface.

Moisture needs can also be estimated by balancing daily water use against daily needs, based upon root volume and water-holding capacity of the soil. For example, nearly all blueberry roots are located in the top 60 cm (2 ft) of soil. A silt loam soil has a typical water-holding capacity of 0.18 in of water per in of soil (Table 4-2). Therefore, 24 in (60 cm) of soil times 0.18 in equals 4.32 in of total water-holding capacity. So that water is readily available to the plant, total available water should not fall below 50% of the total capacity, or 2.16 in. Well-watered, vigorously growing blueberry plants can transpire up to 0.25 in or more of water per day during peak fruiting periods. This is often twice as fast as pan evaporation. At that rate, blueberries growing in a silt loam soil will lower the moisture content to the critical 50% level in about nine days (or

Table 4-2. Water holding capacity of soils with various textures.

Texture	Available water holding capacity (inches water/inch soil)
Sand	0.05
Fine Sand	0.08
Sandy Loam	0.11
Loam	0.16
Silt Loam	0.18
Clay Loam	0.19
Silty Clay	0.20
Clay	0.22

Source: Pritts and Handley 1989.

8.64, to be exact). If significant rainfall does not occur within this period, irrigation is advisable.

Under water stress, plant growth is severely affected. Typical drought symptoms include graying, then reddening of the foliage; weak, thin shoots; early defoliation; and decreased fruit set. Supplemental irrigation should begin before any of these symptoms appears. However, excessive watering can leach nutrients and water-log the soil.

Estimating the size of irrigation ponds is important. To apply 0.25 acre-inch of water per 24-hour period requires 6789 gallons (27,154 gallons per acre-inch × 0.25 acre-inch), or a continuous flow rate of about 4.7 gallons per minute per acr over a 24-hour period (assuming the irrigation system is operating at 100% efficiency). However, if the system is only 75% efficient, as is often the case, then 6.3 gallons per minute must be delivered. If the grower wishes to irrigate only over an eight-hour period, then the system must deliver about 19 gallons per minute per acre. Be sure that the discharge rate does not exceed the restocking rate of the water source (Geohring 1989).

Water can be delivered in several ways, among them furrow, sprinkler, and trickle (drip) irrigation. The furrow system requires level land, a low soil-water intake rate, and the use of raised beds. Most commonly used are the sprinkler and trickle systems.

Sprinkler System

Among the advantages of the sprinkler system is its ease of maintenance, its relatively low cost, and its adaptability for frost protection during bloom (as discussed above). It is not easily used on slopes greater than 10% and it may cause erosion or crusting on some soils. It also uses *more* water *less* efficiently than the trickle system, by watering noncrop portions of the plantation. The sprinkler should not be used during ripening, a most critical time for water consumption, since wetting the fruit surface at this time can crack the fruit, particularly after drought conditions have occurred. Also, researchers have found that terminating the use of overhead sprinklers as much as 30 days before harvest still significantly darkened the fruit of 'Croatan' and 'Harrison,' probably by affecting the waxy bloom (Mainland 1989a).

This system has also been used to increase vegetative growth, by maintaining a cooler leaf temperature, providing high humidity, and decreasing water loss from the plant (Mainland 1989b).

Sprinklers should be spaced so that their watered areas overlap by about 50%. Remember that about 15% of sprinklers' water is lost through evaporation before it even reaches the ground. This percentage increases with increasing temperature and windspeed. Therefore, sprinklers should run about 15% longer than the necessary amount of time calculated.

Trickle Irrigation

The trickle irrigation system, also called drip or microirrigation, is more efficient than the sprinkler system, because it places water more precisely into the root zone of the bush and can partially replace the need for soil organic matter (Korcak 1992). Less water is wasted through evaporation, and the smaller motors and pumps used in this low-pressure system (5-40 psi) result in lower operating costs. Liquid fertilizer can also be fed through this system (fertigation). It is useless, however, in reducing potential frost damage, and the low operating pressure makes the entire system more sensitive to gradients. Further, because of the small orifice size, water must be filtered through a fine, 200-mesh screen to prevent clogging. The grower should use caution to ensure that the entire root system receives adequate water through even distribution of emitters around the plant. Gough (1984a,b) and Abbott and Gough (1986) reported that potted blueberry plants receiving water or liquid fertilizer on only one side grew and fruited only on that side; the non-watered, nonfertilized side did not grow or fruit, and in some cases died. Even watering is essential for best growth and production.

Any type of irrigation system is a long-term investment and can be quite costly. Consult a professional irrigation company for the best method and design.

The quality of the water is an important element of an irrigation system. Sediments and algae contaminants from the well or pond can clog nozzles and emitters. Filters (30-200 mesh) can be used to clear sediments, and periodic water treatment will keep algal growth under control. Water with high sodium, boron, or calcium content; high pH; a salt content exceeding 0.1% total salts; and more than

300 ppm chloride is unsatisfactory for best blueberry growth. Recently, Wright et al. (1992) reported that root and shoot growth were depressed in plants with increasing salinity. Affects of sodium sulfate could be ameliorated by the addition of low concentrations of calcium, though those of sodium chloride could not.

Water can also be supplied to the plantation during the dormant season in the form of snow. The snow melt is held in the soil and adds to the total moisture content, thereby aiding early spring growth. Snowfall can also add some nitrogen to the soil at the rate of about 1 lb per acre-ft (3.4 kg per hectare-meter) of snow; however, this amount is negligible to blueberry production. Snow itself has fair insulation value and can protect the root system and upper portions of the plant from severe cold (Wildung and Sargent 1989a,b). When the air temperature is −14°F (−26°C), the temperature at the snow's surface is −1°F (−18°C) and 16°F (−9°C) under 3 in (7.6 cm) of snow. Under 6 in (15 cm), it is 22°F (−2°C). Compacted snow loses most of its insulating value.

Other forms of water, such as hail, sleet, and ice, can cause considerable mechanical damage to the plant.

Chapter 5

Soil

Soil is a complex, dynamic substance made up of living and non-living things. Minerals, air, water, the decaying remnants of once-living plants and animals, and microorganisms, such as bacteria and fungi, all interact to keep soil in a state of constant change.

Particles of minerals come in different sizes. Clay (< 0.002 mm), silt (0.002-0.05 mm), sand (0.05-2 mm), and gravel (2-4 mm) all make up soils, though good agricultural soils usually contain only the first three in various proportions. The term "loam" refers to a soil containing about 35-40% sand, 35-40% silt, and 10-25% clay. The remainder is pore space occupied by a combination of air and water. If loam has a larger percentage of silt than either sand or clay, it is called a silt loam. Silt loams are considered the best agricultural soils. These mineral soils contain less than 20% organic matter. Those soils containing more than 20% organic matter are called organic soils. A soil with 20-65% organic matter is a muck soil, and one with more than 65% is a peat soil. A natural blueberry soil is a sand or loamy sand (> 70% sand; < 15% clay), with low fertility, a pH of about 5.5, and more than 4% organic matter (Korcak 1989a). Such soil also has a high polyphenol content and specific rhizosphere flora and fauna that may benefit blueberry plant growth (Blasing 1989b). Although blueberries naturally do not grow well on upland, mineral soils, they do tolerate a wide range of soils, particularly if they are light, "fluffy," and well-drained. Trickle irrigation can sometimes replace the beneficial affects of organic matter (Korcak 1992a). Recently, breeding programs have been initiated to better adapt the highbush blueberry to the non-typical upland soils (Erb 1987). Erb et al. (1993) reported that selecting new clones for increased leafiness, high photosynthetic rate, and a more energy efficient root system could lead to improved tolerance of mineral soils. Establishing a planting on improper soils is one of the most common errors among growers around the world. The grower will face a constant stream of problems and the plantation will never yield to its greatest potential. In many cases, plant loss will exceed 50%.

Soils under old pastures generally have good characteristics, and the use of cover and green manure crops can improve the texture even more. Any good loam soil will be suitable for blueberry growth after some amendments, though very sandy or very clayey loams should be avoided. Sandy soils are usually too low in organic matter, drain too quickly, and heat too excessively in the summer for good growth. Clay soils are poorly drained and compacted. These have poor aeration and restrict root growth. Loams with an organic matter content of 3%-15% are excellent. In a recent German study, researchers reported that blueberries generally made poor growth when planted on arable land but good growth when planted on virgin forest soil (Blasing 1989a,b). These differences could not be explained by soil characteristics or nutrient supply. Plants on the forest soil produced sparse, spreading root systems, while those on arable land produced dense root systems with small volume. Mycorrhizal infection was greater in the forest soil.

SOIL REACTION

Coville (1910) first reported that the blueberry plant required an acid soil for best growth and he determined that the optimum pH ranged from 4.3-4.8, although the plants made satisfactory growth in the range of 4-5.2. In soil having a pH below 3.5, aluminum and manganese become very soluble and toxic to the plant, roots become coarse and stubby and cease growing, and marginal leaf scorch appears, followed by stunted growth and death (Merrill 1944; Korcak 1989b). At a soil pH above 5.2, nitrogen is converted from the useful acidifying ammonium form to the less useful nitrate form, and iron chlorosis becomes a problem, since soil iron becomes chemically bound and unavailable to the plant. This results in an iron deficiency, which in turn interferes with chlorophyll synthesis.

The normally green leaves turn yellow, photosynthesis can no longer proceed, and the plant dies. Some researchers suggest that iron-efficient genes may exist that allow progenies of certain crosses to release hydrogen ions from their roots into the soil solution, thereby acidifying it by as much as 1-2 pH units and increasing the availability of iron (Brown and Draper 1980; Leonard 1984). If such genes do exist, they might be incorporated into commercial cultivars and allow plantings on normally unacceptable soils. Austin and Bondari (1992) working with rabbiteye cultivars, reported that total fruit yield and uniformity of ripening was greatest when soil pH was below 5.

Finn et al. (1991) reported that cultivars having a higher percentage of lowbush blueberry (*V. angustifolium*) in their parentage generally had a greater tolerance to higher soil pH. Korcak (1992b) found that high pH, not high soil calcium, was responsible for poor blueberry growth.

Growers should always adjust the pH of their soils before planting, and make periodic checks to be sure it remains in the optimum range for growth.

ORGANIC MATTER

Organic matter is the residue of once-living organisms. In the soil, it helps retain moisture, reduces the leaching of nutrients by holding cations on negative exchange sites, and increases the availability of some micronutrients (especially iron) by acidifying the soil during decomposition. Organic matter provides energy for soil microorganisms, and bacteria digesting the organic particles produce complex carbohydrates that cement soil particles together into aggregates. These increase porosity and "fluffiness" and the soil is said to be "friable," or in good tilth. On somewhat sandy soils, organic matter aids in the retention of potassium, calcium, and magnesium and it holds nitrogen in the useful ammonium form readily absorbed by the plant. It can also reduce stress from aluminum toxicity (Korcak 1988). Blueberry roots are limited in their downward growth by the availability of organic matter in the lower levels of the soils, and the growth and yield of the plant are directly proportional to the total amount of soil organic matter (Shoemaker

1978). Since this soil component is so important for good production, select a soil with at least 3%-5% organic matter content. Although some soil management practices will add small amounts of organic matter to the soil, it is seldom possible to make large, permanent increases after planting.

MOISTURE

Water is held in the soil by three different methods. "Gravitational" water saturates the pore spaces in excessive amounts. It is readily available to the plant, but drains quickly. After this drainage has ceased, usually in a few days, the soil is said to be at "field capacity," and the remaining water available for the plant is more tightly held at higher tensions in the small spaces between soil particles. This is termed "capillary" water and provides the bulk of moisture used in plant growth. When all the capillary water has been extracted from the soil, only "hygroscopic" water remains. This is bound so tightly to the soil particles at tensions greater than 15 atmospheres (15 bars), that it is unavailable for plant use. When half the capillary water supplies are exhausted, the plant will begin to wilt (incipient wilt). The plant has reached the "permanent wilting point" when the capillary water supplies are entirely exhausted, and it will not recover its turgidity even when placed in an atmosphere with 100% relative humidity. Under drought conditions, water cannot be absorbed and neither can nutrients. Stomates close and photosynthesis is reduced. Shoots weaken and become thin and spindly. Leaves turn a bluish gray-green, then red, and finally yellow and necrotic. They often will turn upward against the shoot. Defoliation occurs prematurely, fruit set is decreased, and the remaining fruit do not size. Finally, under severe drought, some shoot tips and whole plants die.

Erb et al. (1991) found that blueberry plants adjusted over time to moderate moisture stress by lowering their stomatal conductance and increasing water-use efficiency. As stomatal conductance decreased, so did transpiration and photosynthesis.

The opposite extreme occurs under flooded conditions. Excessively wet soil can result in substantial winter heaving and root death. Since roots are no longer able to absorb water, the symptoms of flooding often mimic those of drought. These develop over a period of time. When plant death under wet conditions is fairly sudden, usually within ten days, and is preceded by rapid desiccation, browning, and death of the leaves and stems without leaf abscission, the actual cause of death may be *Phytophthora* root rot, not flooding (Crane and Davies 1988). Under flooded conditions, gas exchange between the soil and the air is reduced. Oxygen content, which is present at a level of about 20% in a well-aerated soil, is reduced and quickly used up by microorganisms (Ponnamperuma 1984). Microbial metabolites accumulate and become toxic, as do by-products of anaerobic metabolism in the root itself (Crane and Davies 1989). These include alcohol, methane, methyl compounds, and aldehydes. Some nutrients in the soil become reduced and toxic, and leaching and denitrification deplete soil nitrogen. Because of the death and poor functioning of some roots, stomates close, photosynthesis is reduced, and nutrients are not absorbed because of decreased root permeability (Abbott and Gough 1987b; Crane and Davies 1987) (Figure 5-1a,b). Hormone imbalances occur in the plant, as does leaf epinasty, chlorosis, and abscission. New aerenchyma tissue is formed in the plant to facilitate movement of oxygen from shoots to roots (Abbott and Gough 1987c) (Figure 5-2a,b). Wet soils warm much more slowly in the spring, requiring five times greater heat input than dry soils to raise their temperature 2°F (1°C). This further retards growth.

Plant response to flooding depends upon the season and length of the flooding period. Abbott and Gough (1987b,d) reported that young 'Bluecrop' plants survived 30 months of continuous flooding under standing water, but that vegetative growth was substantially depressed after about four months. Flooding begun in April was more damaging than flooding begun in August and December. Unfortunately, this is the time during which most field flooding is apt to occur. Response is probably related to the plant's developmental stage at the time of flooding. Roots died faster if flooded during their normally active growth period, and more slowly if flooded during periods of less activity. Flooded plants developed about 70% fewer flower buds and 60% fewer flowers per bud. Bloom was delayed about six days, fruit set decreased 45%, and premature fruit drop increased. Weight, diameter, and flavor of individual fruit were

Figure 5-1a,b. Scanning electron photomicrograph of a cross-section of a 'Blue-crop' root. Non-flooded plant, scale bar = 11 μm (a); flooded plant, scale bar = 10 μm (b). Epidermis (E); cortex (C); endodermis (Ed); 2nd stele (S). Note enlarged epidermal cells and crushed stellar tissue in roots from flooded plants. (Source: Abbott and Gough 1987b).

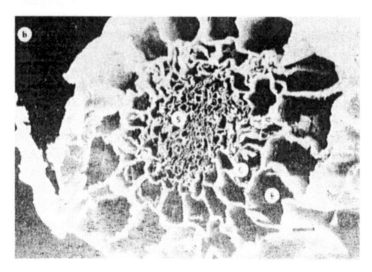

Figure 5-2a,b. Light photomicrograph of a cross-section of a 'Bluecrop' stem. Non-flooded plant, scale bar = 65 μm (a); flooded plant, scale bar = 69 μm (b). Epidermis (E); cortex (C); and aerenchyma (A). (Source: Abbott and Gough 1987b).

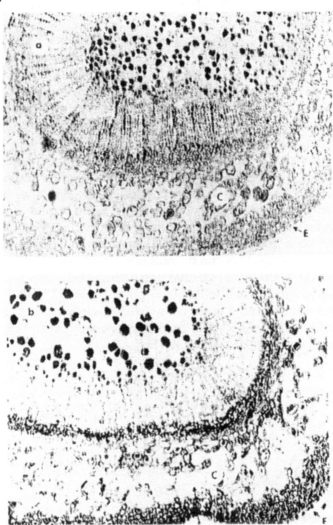

reduced significantly by flooding. Davies and Flore (1987) found that photosynthesis was significantly reduced after only one to two days of flooding and that respiratory use of carbon exceeded synthesis after ten to fourteen days (i.e., the plant was beginning to live on limited reserves of food).

Soil drainage is a very important consideration when selecting future planting sites. In spite of popular belief, blueberries *do not* prefer wet soils and will not thrive in such conditions. Observe the soil in, and somewhat below, the future root zone. Well-drained soils will be a uniform, bright color–red, yellow, or brown. Mottling or gray colors indicate repeated short-term flooding. Soils with a uniform, light-gray color indicate prolonged periods of flooding. If your site inspection suggests flooding, consider implementing surface drainage. This involves grading to a gentle slope and possibly construction of a system of dikes and ditches to direct water flow. Subsurface drainage involves laying a system of buried drains and can be quite costly. In such situations, choosing an alternate planting site may be a more cost-effective approach. Under emergency conditions, drainage tiles can be backflooded and used for irrigating the plants (D. Hartmann, personal communication).

The water table should range in depth from 14-32 in (31-81 cm) below the soil surface. Higher tables will severely restrict root system development, while lower ones may not provide sufficient water to the plant under drought conditions.

Chapter 6

Site Selection

Good site selection is one of the most neglected steps in blueberry production. It can save considerable expense later on and mean the difference between profit and loss. Once the grower has determined that highbush blueberries will grow well in the general area, specific planting sites should be selected. Wild plants growing in the area–such as huckleberry, laurel, and wild blueberry–or a mixture of pine (*Pinus sp.*), red maple (*Acer rubrum*), and white cedar (*Chamaecyparis thyoides*) generally indicate suitable soil. The local climate of the site should be studied for at least a couple of seasons prior to planting, and the following conditions considered.

Slope. Planting on slopes is desirable because they allow heavy, cold air to flow down and away from plants. Temperatures on the slope remain very uniform within the bush canopy. Because cold air drains to the lowest point, keep bushes at least 20-25 yd (m) above the base of the slope. Plant later-blooming cultivars toward the base and earlier-blooming ones near the top (Table 6-1). A difference in elevation of as little as 100 ft (30 m) can mean a difference in temperature of 10°F (6°C). Planting at the very top of a slope exposes plants to excessively windy, dry conditions that restrict blueberry plant growth. If prevailing winds are strong and persistent, locate the planting on the leeward side.

The grower will rarely have a choice of whether or not to plant on a slope, and even more rarely be able to choose the direction of a slope. But, in the event that there are several choices, the following should be considered. In the northern hemisphere, north slopes can be 5°-10°F (2.75°-5.5°C) colder than south slopes during spring and fall. As a result, spring bud growth is delayed and late frost damage can be reduced. However, harvest may also be slightly delayed. A

Table 6-1. Average characteristics of bloom at Kingston, Rhode Island, U.S.A.

Cultivar	Days with open flowers	Length of full bloom[z]	Date of full bloom[y]
Earliblue	27	4	9
Bluetta	30	5	11
Northland	27	2	17
Patriot	20	3	17
Bluehaven	27	2	20
Coville	25	1	20
Darrow	25	2	20
Elizabeth	23	2	21
Herbert	16	2	21
Berkeley	23	2	22
Bluecrop	25	2	22
Blueray	25	2	22
Collins	27	4	22
Lateblue	21	1	23
Elliott	18	1	24

[z]Recorded in days. Full bloom is defined as the point in time when the first corollas begin to fall.
[y]May
Source: Gough et al. 1983; R. E. Gough, unpublished data.

south slope warms faster in the spring and buds develop more quickly, rendering them more susceptible to spring frost damage. This is particularly true for the lower half-highs. The soil may be poorer than on other slopes and it may dry more quickly, resulting in early, smaller fruit (though the growing season may be up to 30 days longer than on north slopes). Warming late-winter sun can cause some damage to plants on south slopes. The features of east and west slopes lie between those of north and south. West slopes are generally warmer than east slopes, but are often exposed to prevailing winds. In far northern areas, planting on north-facing slopes is generally the safest.

Steep slopes not only increase the potential for erosion damage but make it difficult to manage equipment. Driving equipment on

15°-20° slopes is hazardous, and planting in dense patterns across the slope to reduce soil erosion can impede air drainage, resulting in the potential for greater spring frost damage.

Sunlight. The importance of light has been stressed in Chapter 3. Shutak (1966) reported that although the crop is little affected when sunlight is reduced by 30%, greater reductions in intensity result in formation of fewer flower buds and later-ripening fruit. Where bushes are shaded by trees, both light reduction and root competition become factors in production. Therefore, locate the plantation where it receives full sun throughout the day.

Wind. Wind turbulence increases with the distance above the soil surface. There is very little wind within plant rows and hedgerow plantings. Canes and branches that touch or interlock can produce severe air stagnation and increase potential for damage from both frost and diseases, as well as interfere with flower bud formation and the distribution of spray materials. Wider spacings increase air turbulence and circulation, lower the relative humidity within the canopy, and increase the exchange of carbon dioxide between the air and leaf surface. Where strong winds predominate, the grower should reconsider the advisability of planting blueberries. If it is still deemed feasible, windbreaks of densely growing species that have branches all the way to the ground, such as spruce (*Picea sp.*) and arborvitae (*Thuja occidentalis*), may be necessary. A windbreak 35-ft (10-m) tall will give good protection to a plantation out to 400 ft (120 m) and it will provide some protection to about 1500 ft (450 m) to leeward (Janick et al. 1981). Such breaks, however, can result in frost pockets during temperature inversion, by impeding air circulation and by reducing heat loss from the plant. Still, they also reduce the amount of mechanical wind damage and increase the amount of snow insulating plants from extreme cold. Always allow for some porosity in windbreaks and see that they reduce windspeed by no more than about 50%, to limit the amount of stagnant air. Running breaks across slopes could impede the downhill flow of cold air and increase the potential for frost damage. Considering all factors, the cost of establishing a windbreak is not usually justified unless the potential loss from wind damage would be substantial.

Bodies of Water. Because of its great specific heat, and latent heat of fusion, water undergoes a smaller temperature fluctuation than either soil or cover crops and can influence plantation air temperatures considerably. Plants near large bodies of water, such as lakes, bays, and the ocean, are cooled in spring by air passing over the cold water. They therefore bloom later and are less susceptible to spring frost damage. Likewise, air passing over warm water in the fall will keep the plantation warm and allow late cultivars to ripen their fruit before frost. This influence decreases substantially as distance from the water increases. Because weather patterns generally move from west to east in North America, the eastern shore is often the best location for planting.

Chapter 7

Plant Selection

Purchasing high-quality stock is the grower's wisest and least expensive investment. Cutting corners here could lead to disaster later. Always purchase healthy, vigorous stock that is certified disease-free and true-to-type. Plants may be one-year-old rooted cuttings or larger plants two to five years of age.

Rooted cuttings are less expensive initially, but they require great care if field-planted directly and are better grown in small beds or containers for a year. Beds with soil rich in organic matter are easily constructed and the cuttings are lined out in rows raised 6-8 in (15-20 cm) high, spaced about 18 in (45 cm) apart. Cuttings are spaced 6-10 in (15-25 cm) apart within rows. The raised rows provide for better drainage and root development. Beds should be mulched to reduce water loss and help control weeds. A light, non-compacting mulch, such as sawdust or woodchips, is satisfactory. Supplemental irrigation may also be necessary to keep the plants growing vigorously. Adequate moisture levels and weed control are the greatest problems facing the grower of rooted cuttings.

Cuttings may also be grown in containers. This often results in less transplant shock later on. However, plants with pot-bound root systems recover very slowly when field-set and may lag several years behind other plants. Remove blossoms on all one-year plants, whether bed- or container-grown.

It is usually best to purchase dormant, two- to three-year-old plants that are 12-24 in (30-60 cm) high. Plants grown in nurseries in the southern United States may reach this size in about 19 months. The higher cost of plants older than this is not justified since after several years in the field, there is very little difference in the size of the two groups. Also, the older, larger plants may suffer more transplant shock, and mortality may be higher than with younger plants.

Balled and burlapped (B&B), canned (potted), or bare-rooted plants are available. In most areas, the first two are preferred, since they can be planted with least disturbance of the root system. However, in Florida bare-root plants usually outgrow potted plants because of their much larger root systems. (P. Lyrene, personal communication). Bare-rooted plants are often shipped in plastic, which should be removed upon arrival, and the roots are covered with damp sawdust or peat moss to reduce desiccation.

It is generally safer to purchase plants from nearby nurseries to avoid a delay in planting, though nurseries specializing in blueberries offer a wider selection of cultivars and often have lower prices. Be sure to order plants at least six months before planting, to assure the greatest probability of getting the cultivars of your choice.

CULTIVAR SELECTION

The term "cultivar" refers to the common name of a kind of plant, replacing the older term "variety." As mentioned earlier, the term is derived from the first syllables of the words "*culti*vated *vari*ety," and it is designated by single quotation marks in scientific literature–as for example, 'McIntosh' apple, 'Marglobe' tomato, or 'Earliblue' blueberry.

There are dozens of cultivars of highbush blueberry, each with its own peculiarities that make it more or less adapted to a particular region of the world. Recently, new, low-chill highbush blueberries have been developed through the hybridization of southern (*V. darrowi, V. ashei*) and northern (*V. corymbosum*) species, reducing the normal chilling requirement from about 800 hours to 150 hours (Table 7-1). This has allowed planting of blueberries in Florida and the warmer parts of the world. However, such cultivars are not particularly winter-hardy and should not be planted in areas having severely cold winters.

Another significant extension of climatic adaptation has come through the development of half-highs. These cultivars are the product of interspecific hybridization of lowbush and highbush plants,

Table 7-1. Chilling requirements for some southern highbush cultivars.

Cultivar	Chilling Hours
Avonblue	300-400
Blue Ridge	500-600
Cape Fear	500-600
Misty	150
Cooper	400-500
Flordablue	150-300
Georgiagem	350
Gulf Coast	400-500
O'Neal	200-500
Sharpblue	150-300
Sunshine Blue	150

resulting in progenies about half the height of typical highbush. This reduced stature results in better survival in cold areas, where heavy snow covers and insulates the plants. Other new cultivars have a greater disease resistance, allowing plantings to succeed in areas formerly considered marginal.

To provide a longer harvest season and larger, better-flavored fruit, select early, mid-season, and late-ripening cultivars. Most older cultivars still have some merit, but they have generally been superseded by newer ones and are not listed in this book. Because of the large selection available, and because of varying local climates within each region, prospective growers should consult with their Cooperative Extension Service, agricultural research stations, and local growers before making their selections.

A detailed description of each cultivar follows, along with parentage, year of introduction, and region of adaptation, based upon the USDA hardiness zone map. Also, citations are given for the most recently introduced cultivars.

CULTIVAR DESCRIPTIONS

Avonblue: 'Florida 1-3' × ['Berkeley' × ('Pioneer' × 'Wareham')]. 1977. 9b-7a.

Bush: Small, spreading.
Fruit: Medium-large, light blue, small scar, very high quality.
Notes: Requires heavy pruning. Prune right after flowering to prevent overfruiting. Self-fruitful; requires 300 chilling hours; ripens May 7 at Gainesville, Florida; low plant vigor limits it to the best soils.

Berkeley: 'Stanley' × ('Jersey' × 'Pioneer'). 1949. 7a-5a.
Bush: Upright, vigorous, open-spreading.
Fruit: Medium-large, light blue, fair quality, large scar.
Resistance: Cracking.
Notes: Excessive drop with delayed harvest; inconsistent productivity; yield better on heavier soil; lightest blue of all; very susceptible to stem canker in North Carolina; canes may break during mechanical harvesting; ripens July 25 at South Haven, Michigan.

Bluechip: 'Croatan' × 'US11-93.' 1979. 7b-6a.
Bush: Vigorous, upright.
Fruit: Very large, excellent color and quality.
Resistance: Botryosphaeria corticis canker, mummyberry, *Phytophthora* root rot, bud mite.
Notes: Easy to prune; very susceptible to stem blight in North Carolina; ripens June 15 at Castle Hayne, North Carolina.

Bluecrop: ('Jersey' × 'Pioneer') × ('Stanley' × 'June'). 1952. 7a-4a.
Bush: Upright, moderately vigorous, productive.
Fruit: Large, very light blue, medium quality, small scar.
Resistance: Cracking, drought.
Notes: Consistently productive. May have excessive fruit/leaf ratio; susceptible to stem canker in North Carolina; adapted to mechanical harvest; ripens July 15 at South Haven, Michigan. This is the leading cultivar in the world in terms of acreage. "Redberry," the condition where berries remain red and do not fully color, can be troublesome in this cultivar. Vorsa (1991) attributes this to overcropping, lack of fertilization, inadequate detailed pruning, and poor pollination. He also suggests that because this cultivar is nearly 60 years old (seed to present), it may simply be "running out." In a recent study, the parentage of this cultivar has been questionable. Haghighi

and Hancock (1992) reported that 'Bluecrop' may not contain 'Rubel' cytoplasm as originally thought. 'Rubel' occurs in the parentage of 'Jersey,' 'Stanley,' and 'June.'

Blue Gold: 'Bluehaven' × Me-US5 ('Ashworth' × 'Bluecrop'). 1989. 7a-4a.
Bush: Vigorous, low-growing, highly productive.
Fruit: Medium size, with good flavor, color, and firmness. Scar is small and dry. Ripens late, with 'Jersey.'
Notes: Highly productive, late-season fruit suitable for commercial packing and U-Pick. Produces an overabundance of branches and flower buds, requiring good pruning to force upright growth and reduce crop load. Ripens July 30 at South Haven, Michigan. (Draper, personal communication).

Bluehaven: 'Berkeley' × (lowbush × 'Pioneer' seedling). 1968. 7a-5a.
Bush: Upright.
Fruit: Large, light blue, very small dry scar, excellent quality.
Notes: Holds quality well; good ornamental value; concentrated ripening; ripens July 12 at South Haven, Michigan.

Bluejay: 'Berkeley' × 'Michigan 241' ('Pioneer' × 'Taylor'). 1978. 7b-4b.
Bush: Vigorous, upright, slightly spreading.
Fruit: Medium-large, light blue, mild flavor, small scar.
Resistance: Premature drop and cracking.
Notes: Long stems help fruit removal with vibration; good ornamental value; adapted to mechanical harvesting, with 70% of the fruit picked at the first harvest; ripens July 10 at South Haven, Michigan. 'Bluejay' blooms late, potentially avoiding late frost, but ripens its fruit early (Hancock et al. 1991).

Blueray: ('Jersey' × 'Pioneer') × ('Stanley' × 'June'). 1955. 7a-4a.
Bush: Upright, very vigorous, spreading, productive.
Fruit: Very large, light blue, firm, excellent quality, medium scar.
Resistance: Cracking.
Notes: Blossoms typically pink-ribbed; leading cultivar for U-Pick;

some tolerance to stem canker in North Carolina, but susceptible to stem blight; ripens July 18 at South Haven, Michigan.

Blue Ridge: 'Patriot' × US-74 (Fla 4-Bx 'Bluecrop'). 1987. 9a-7b.
Bush: Vigorous, very upright, productive.
Fruit: Large to very large, very light blue, tart, firm; wet scar similar to 'Blueray.'
Notes: Susceptible to stem canker (*B. corticis*) and mummyberry; tolerant to stem blight (*B. dothidea*); suitable for U-Pick and local sales in Coastal Plain, Piedmont, and lower mountains of the southeastern United States. Broadly adapted in North Carolina; requires 500-600 chilling hours; ripens May 15-25 in south Georgia (Ballington et al. 1990a).

Bluetta: ('North Sedgewick' lowbush × 'Coville') × ('Earliblue'). 1968. 7a-5a.
Bush: Short, compact-spreading, moderately vigorous, productive.
Fruit: Small to medium, light blue, firm, broad scar.
Notes: Difficult to prune, softens rapidly after ripening, replacement for 'Earliblue' and 'Weymouth'; not attractive in fresh pack; ripens July 1 at South Haven, Michigan.

Bounty: 'Murphy' × G-125 (F-72 ['Wareham' × 'Pioneer'] × E-7 ['Berkeley' × 'Earliblue']). 1988. 9a-7b.
Bush: Moderately vigorous, spreading habit. Requires extensive early training to develop semi-upright plant.
Fruit: Very large, excellent flavor, scar and color. Firm.
Notes: Susceptible to mummyberry, anthracnose fruit rot, and blueberry bud mite; some field tolerance to stem blight and stem canker. Plant on soils high in organic matter. Ripens with 'Murphy' and has extended shelf-life. Intended for North Carolina production in areas with cane canker (*B. corticis*) and stem blight (*B. dothidea*) problems (Ballington 1989; Ballington et al. 1989b).

Burlington: 'Rubel' × 'Pioneer.' 1939. 7a-5a.
Bush: Slightly spreading, wide crown.
Fruit: Medium small, very firm, very small dry scar, excellent shipping quality.

Notes: Not consistently productive in colder climates; adapted to mechanical harvesting; ripens August 24 at South Haven, Michigan.

Cape Fear: US-75 (Fla. 4B × 'Bluecrop') × 'Patriot.' 1987. 9a-7b.
Bush: Vigorous, semi-upright, very productive.
Fruit: Large to very large, light blue, small scar (better than 'Croatan,' but worse than 'Bluechip' and 'Bounty'); excellent flavor and firmness. High removal force. Develops a metallic flavor if allowed to overripen in warm weather.
Resistance: Tolerant to stem blight.
Notes: Susceptible to stem canker. Ripens May 15 in south Georgia and with 'Croatan' in North Carolina. Recommended for North Carolina and the Coastal Plain, Piedmont, and low mountains of the southeastern United States. Requires 500-600 chilling hours (Ballington et al. 1990a).

Challenger: Obsolete name. See 'Misty.'

Collins: 'Stanley' × 'Weymouth.' 1959. 7a-5b.
Bush: Erect, vigorous, fairly productive.
Fruit: Medium to large, light blue, firm, highly flavored.
Resistance: Cracking, premature drop.
Notes: May not sucker freely; ripens early mid-season.

Cooper: G-180 [G-100 ('Ivanhoe' × 'Earliblue') × 'Collins'] × US75 [Fla. 4b (*Vaccinium darrowi*) × 'Bluecrop']. 1987. 9a-7b.
Bush: Moderately vigorous, upright, moderately productive.
Fruit: Medium-sized (slightly larger than 'Gulf Coast'), good flavor, good color, firm, good scar.
Notes: One of the first two low-chilling highbush cultivars released from the USDA-ARS (Agricultural Research Service) breeding program at Poplarville, Mississippi. Outstanding in earliness and superior fruit quality. Flowers later than 'Climax.' Ripens May 15 in south Georgia, or about two weeks before 'Climax.' Recommended for Coastal Plain, Piedmont, and the low mountains of the southeastern United States. Highly susceptible to stem canker. Requires 400-500 chilling hours (Draper, personal communication).

Coville: ('Jersey' × 'Pioneer') × 'Stanley.' 1949. 7a-5b.
Bush: Vigorous, upright spreading, very productive.
Fruit: Very large, medium blue, excellent quality, small scar.
Resistance: Preharvest drop, cracking.
Notes: Self-incompatible, flowers unattractive to bees, abnormal pollen development, dehiscence, and germination; do not plant in solid blocks. Adapted to mechanical harvest. Ripens August 1 at South Haven, Michigan. Very susceptible to stem canker in the Coastal Plain of the southeastern United States.

Croatan: 'Weymouth' × ('Stanley' × 'Crabbe 4'). 1954. 8a-6b.
Bush: Upright, vigorous.
Fruit: Large, dark blue, fair quality.
Resistance: Bud mite, stem canker, cracking.
Notes: Ripens early, with 'Weymouth'; very susceptible to stem blight in North Carolina.

Darrow: ('Wareham' × 'Pioneer') × 'Bluecrop.' 1965. 7a-5b.
Bush: Upright, vigorous.
Fruit: Very large, firm, light blue, very good quality, small scar.
Resistance: Cracking.
Notes: May lack hardiness in some areas, inconsistent productivity. Ripens late, with 'Coville.'

Dixi: ('Jersey' × 'Pioneer') × 'Stanley.' 1936. 7a-5b.
Bush: Open-spreading, productive.
Fruit: Very large, medium blue, large scar, very high quality.
Notes: Cracks easily. Ripens late, after 'Jersey.'

Duke: ('Ivanhoe' × 'Earliblue') × 192-8 (E-30×E-11). 1985. 7a-5a.
Bush: Vigorous, upright.
Fruit: Medium size, light blue, small dry scar, firm. Ripens slightly later than 'Bluetta.' Flavor is mild, but becomes more aromatic after several hours in cold storage.
Notes: Alternative to 'Bluetta.' Relatively late flowering. May be strongly self-fruitful. Numerous canes are stocky, but moderately branched, allowing sunlight to penetrate the canopy. Buds and wood tolerate fluctuating winter temperatures very well. Susceptible to

stem canker in North Carolina. Ripens July 10 at South Haven, Michigan. Harvest completed in two pickings (Draper, personal communication).

Earliblue: 'Stanley' × 'Weymouth.' 1952. 7a-4b.
Bush: Hardy, upright-spreading, productive.
Fruit: Large, aromatic, light blue, good quality, medium scar.
Resistance: Cracking, premature drop.
Notes: Erratic fruit set, very susceptible to phomopsis canker. Ripens July 1 at South Haven, Michigan. No longer recommended for commercial planting in some areas.

Elliott: 'Burlington' × ['Dixi' × ('Jersey' × 'Pioneer')]. 1973. 7a-4a.
Bush: Upright, very hardy.
Fruit: Medium light blue, good quality, very small scar.
Resistance: Mummyberry.
Notes: Tart until 60% of fruit are ripe; ripens extremely late (a week after 'Lateblue,' or about September 1 at South Haven, Michigan). Very tight clusters. Cross-pollination will speed ripening. Adapted to mechanical harvesting, with 70-80% of the fruit picked in one harvest.

Elizabeth: ('Katharine' × 'Jersey') × 'Scammell.' 1966.
Bush: Upright, spreading.
Fruit: Very large, medium blue, fair quality, ships well.
Notes: Inconsistent productivity. Ripens mid-season.

Flordablue: 'Florida 63-20' × 'Florida 63-12.' 1976. 9a-7b.
Bush: Moderately vigorous.
Fruit: Large, light blue, medium firm, medium scar, good quality.
Resistance: Botryosphaeria corticis stem canker.
Notes: Poor shipper; very difficult to root; very susceptible to *Phytophthora*; OK for local markets and U-Pick. Adapted to mechanical harvest. Requires 150 chilling hours. Ripens April 27 at Gainesville, Florida, if pollinated with 'Sharpblue' or 'Misty.'

Friendship: Open-pollinated wild selection from Wisconsin; a suspected natural hybrid of *V. angustifolium* and *V. corymbosum.* 1990. 7a-3a.

Bush: Low (24 in or 60 cm).
Fruit: Small to medium, dark blue, very good "wild" flavor.
Notes: Very hardy half-high. Moderately susceptible to *Godronia* stem canker. Ripens July 8 at South Haven, Michigan (Stang et al. 1990).

Georgiagem: G-132 (E-118 × 'Bluecrop') × US-75 (*V. darrowi* clone Fla. 4-B × 'Bluecrop'). 1986. 9a-7b.
Bush: Upright, tall, vigorous.
Fruit: Medium, small dry scar, very good flavor, firm.
Notes: May be susceptible to stem blight and canker. Plant in well-drained soil, and with early-blooming rabbiteye or another southern highbush for cross-pollination. Requires at least 350 chilling hours. Ripens May 8 at Gainesville, Florida (Austin and Draper 1987).

Gulf Coast: G-180[('Ivanhoe' × 'Earliblue') × 'Collins'] × US75 (*V. darrowi* clone Fla. 4B × 'Bluecrop'). 1987. 9a-7b.
Bush: Vigorous, semi-upright, moderately productive.
Fruit: Medium, small scar, good flavor. Stem often remains attached after harvest.
Notes: Susceptible to stem canker at Gainesville, Florida. Requires 200-300 chilling hours. Ripens May 15 in south Georgia (Draper, personal communication).

Harrison: 'Croatan' × 'US11-93.' 1974. 8a-6b.
Bush: Semi-upright, vigorous.
Fruit: Large, good quality.
Resistance: Cane canker, bud mite.
Notes: Self-fruitful. Ripens early mid-season, after 'Croatan'; quite susceptible to fruit rots; no longer recommended for commercial plantings in North Carolina (Ballington 1989).

Herbert: 'Stanley' × ('Jersey' × 'Pioneer'). 1952. 7a-4a.
Bush: Open-spreading, vigorous.
Fruit: Very large, medium blue, excellent flavor, large, leaky scar.
Resistance: Cracking.
Notes: Inconsistent productivity in cold climates. Adapted to mechanical harvest. Ripens July 15 at South Haven, Michigan. No longer recommended for commercial use in some areas.

Jersey: 'Rubel' × 'Grover.' 1928. 7a-4a.
Bush: Upright, spreading.
Fruit: Small to medium, small scar, excellent flavor.
Notes: Can be harvested mechanically. Foremost berry for processing. Some field tolerance to stem canker in North Carolina. Produces less pollen per flower than other cultivars, and its flowers are less attractive to bees. Without cross-pollination, this cultivar will produce small, late, seedless berries. Ripens July 30 at South Haven, Michigan. A major cultivar in Michigan. Recently, Haghighi and Hancock (1992) have questioned the presence of 'Rubel' in the parentage of 'Jersey.'

Lateblue: 'Herbert' × 'Coville.' 1967. 7a-4a.
Bush: Vigorous, erect.
Fruit: Medium large, light blue, good quality, small scar.
Notes: Excessive stemminess under high harvest temperatures. Adapted to mechanical harvesting. Ripens very late, just before 'Elliott.'

Meader: 'Earliblue' × 'Bluecrop.' 1971. 7a-4a.
Bush: Upright, open-spreading, very hardy.
Fruit: Medium large, medium blue, fair quality, small scar, remains very firm.
Resistance: Cracking.
Notes: May overbear; requires heavy pruning; adapted to mechanical harvesting, with 60-70% of the fruit ripening at once. Suitable where temperatures fall below –25°F (–32°C), with heavy snow. Ripens July 15 at South Haven, Michigan.

Misty: Fla 67-1 (E-30['Berkeley' × 'Earliblue'] × Fla 61-7) × 'Avonblue.' 1989. 9b-7a.
Bush: Tall, upright, small crown.
Fruit: Attractive, light blue, large, firm, small scar.
Notes: Good for low-chill areas; requires 150 chilling hours. Good pollinator for 'Sharpblue.' Easy to root. Susceptible to stem blight. Ripens May 2 at Gainesville, Florida (Lyrene, personal communication).

Morrow: 'Angola' × 'Adams.' 1964. 8a-6b.
Bush: Semi-upright, slow growing.
Fruit: Large, light blue, high quality.
Notes: Short season; replaces 'Angola' in North Carolina.

Murphy: 'Weymouth' × ('Stanley' × 'Crabbe 4'). 1950. 8a-6b.
Bush: Vigorous, spreading.
Fruit: Large, dark blue, fair quality.
Resistance: Stem canker; tolerant to stem blight.
Notes: Slow to come into bearing. Ripens early, just after 'Weymouth.' Old standard cultivar in North Carolina.

Nelson: 'Bluecrop' × G-107(F72 × 'Berkeley'). 1988. 7a-4a.
Bush: Vigorous, medium tall, highly productive.
Fruit: Size, firmness, scar, and color all good and similar to 'Spartan.'
Notes: Highly productive. Ripens late mid-season, with 'Berkeley' or about July 25 at South Haven, Michigan; suitable for commercial packing and U-Pick (Draper, personal communication).

Northblue: B-10 (G-65 × 'Ashworth') × US-3 ('Dixi' × Michigan Lowbush No. 1). 1983. 7a-3a.
Bush: Very low (2 ft or 60 cm), vigorous.
Fruit: Large, dark blue, good quality.
Resistance: Tolerates –20°F (–30°C).
Notes: Self-fruitful; non-rhizomatous; excellent for extreme northern areas. May be susceptible to *Phythophthora* on poorly drained soils. Ripens one to two weeks earlier than 'Bluecrop,' but blooms one week later than 'Bluecrop' (Luby et al. 1986). Fruit are more acidic than those of 'Northsky' and 'Northcountry' and yields are higher (Luby 1991). Though 'Northblue' will set fruit parthenocarpically, cross-pollination is advised for best production (Harrison et al. 1990).

Northcountry: B6 × R2P4. 1986. 7a-3a.
Bush: Moderately vigorous, short (under 40 in or 100 cm).
Fruit: Medium size, very light blue, mild flavor, somewhat soft.
Resistance: Very cold tolerant, to –35°F.
Notes: More productive than 'Northsky,' but less than 'Northblue.'

Self-fruitful. Ripens a week earlier than 'Northblue' and 'Northsky,' or about July 10 at South Haven, Michigan. Harvest by hand or rake (Luby et al. 1986). This cultivar will set very few fruit after self-pollination (Luby 1991).

Northland: 'Berkeley' × (lowbush × 'Pioneer' seedling). 1967. 7a-3b.
Bush: Short (48 in or 120 cm), spreading, canes flexible under snow.
Fruit: Small to medium, medium blue, good quality, small scar.
Notes: Difficult to prune, suckers freely and spreads out of rows in warmer areas. Can be mechanically harvested. Ripens early mid-season.

Northsky: B6 × R2P4. 1983. 7a-3a.
Bush: Extremely low (10-18 in or 25-45 cm), dense, moderately productive.
Fruit: Medium size, light blue, good flavor, somewhat soft.
Resistance: Tolerates –40°F (–40°C) with snow protection.
Notes: Recommended for home gardens. May be harvested by hand or rake. Ripens July 15 at South Haven, Michigan (Luby et al. 1986). 'Northsky' sets few fruit following self-pollination (Luby 1991).

O'Neal: 'Wolcott' × Fla. 64-15. 1987. 9a-7b.
Bush: Vigorous, semi-upright, extended bloom period.
Fruit: Very large, medium blue, very firm, small scar, excellent flavor.
Resistance: Tolerant to some stem cankers
Notes: Susceptible to stem blight. Requires at least 400-500 chilling hours. Blooms early and is susceptible to frost. Extended bloom period in southeastern North Carolina. Ripens early May in south Georgia (Draper, personal communication).

Patriot: ('Dixi' × 'Mich. LB1') × 'Earliblue.' 1976. 7a-3b.
Bush: Low (48 in or 120 cm), open-spreading, very hardy (–20°F or –29°C).
Fruit: Large, light blue, very good quality, small scar.
Resistance: Phytophthora root rot.

Notes: Adapted to home and local production in the northeastern United States. Susceptible to stem canker in North Carolina. Ripens July 5 at South Haven, Michigan.

Rancocas: ('Brooks' × 'Russell') × 'Rubel.' 1926. 7a-3b.
Bush: Upright-spreading, high yielding.
Fruit: Medium, large scar, cracks when ripe, poor quality.
Notes: Up to 30% of the embryo sacs degenerate within three days following bloom. Must be pollinated rapidly for best set. Harvested mechanically for processing only. Ripens July 18 at South Haven, Michigan.

Reveille: NC 1171 (G-111 × Fla 61-7) × NC SF-12-L ('Ivanhoe' × NC297). 1990. 8a-6b.
Bush: Very upright, narrow, vigorous on light soils.
Fruit: Good scar, flavor, high yield, fully self-fertile, medium size and lighter blue than 'Croatan.'
Resistance: Race 4 stem canker (*B. corticis*), possibly race 1 also. Reasonably tolerant to stem blight (*B. dothidea*).
Notes: Fully self-fertile. Broadly adapted to North Carolina soils but may be too vigorous on organic soils. Suitable for mechanical harvesting. Bloom time similar to 'Croatan' and 'Cape Fear,' making frost protection necessary. Fruit removal force very similar to 'Croatan.' Other qualities generally superior to 'Croatan.' Fruit may crack in light-cropping years (Ballington et al. 1990b).

Rubel: Wild selection. 1911. 7a-4a.
Bush: Erect, very productive.
Fruit: Small, light blue, firm, good quality, medium scar.
Notes: Stemmy during drought or if harvest is delayed. Adapted to mechanical harvesting, mostly for processing trade. Ripens July 27 at South Haven, Michigan.

Sharpblue: 'Florida 61-5' × 'Florida 62-4.' 1976. 10a-7b.
Bush: Similar to 'Flordablue,' fast growing, vigorous.
Fruit: Dark blue, similar to 'Flordablue,' small scar. Skin sometimes tears at harvest.
Notes: Fruit should be shipped the day of harvest for best quality. Number 1 southern highbush in Florida. Requires 150 chilling

hours. Ripens April 27 at Gainesville, Florida, if pollinated with 'Flordablue' or 'Misty' (Lyrene and Sherman 1992).

Sierra: US169 (US79 × US79 [Fla. 4B × US-56 (*V. constablaei* × *V. ashei*)]) × G-156 ('Earliblue' × G-77['Coville' × US 11-93]). 1988. 7a-4a.
Bush: Vigorous, upright, productive.
Fruit: Medium size, with small, dry scar, good color, flavor and firmness.
Notes: Highly productive. Recommended as an alternative to 'Toro' and 'Bluecrop.' Because it is an interspecific hybrid of four species, its area of adaptation is unpredictable. Recommended for trial only at this time, especially in colder production areas. Ripens July 10 at South Haven, Michigan. For fresh market and U-Pick (Draper, personal communication).

Stanley: 'Katharine' × 'Rubel.' 1930. 7a-5a.
Bush: Upright, vigorous, few canes.
Fruit: Medium to small, firm, very aromatic, poor scar, excellent flavor.
Resistance: Cracking.
Notes: Easy to prune. Ripens early mid-season.

Spartan: 'Earliblue' × 'US 11-93.' 1978. 7a-5a.
Bush: Vigorous.
Fruit: Medium size, light blue, excellent flavor, firm, dry scar.
Resistance: Mummyberry.
Notes: Berry size decreases considerably after second harvest. Adapted to mechanical harvest, hand harvest, and U-Pick. Bush turns chlorotic in the field and must have well-drained soil, good pH, and proper fertilizer. Very susceptible to stem canker in North Carolina. Ripens July 7 at South Haven, Michigan. 'Spartan' blooms relatively late but ripens its fruit early, potentially escaping late frosts (Hancock et al. 1991; Korcak 1992a).

St. Cloud: Formerly known as MN 167. [B19A (G 65 × 'Ashworth') × US 3 ('Dixi' × Michigan Lowbush #1)]. 7a-3a.
Bush: Half-high. More upright than 'Northblue,' 'Northsky,' or 'Northcountry' and therefore more predisposed to winter injury. Mature height and diameter about 4 ft. (1.3 m).

Fruit: Medium light blue, large, slightly flattened. Similar in size to 'Northland' and 'Bluetta,' smaller than 'Northblue' and 'Patriot' but larger than 'Northsky' and 'Northcountry.' Scar small and dry. Firmer and better flavored than 'Northblue.' Ripens about four to six days ahead of 'Northblue,' or with 'Bluetta.'

Notes: Fruit yield similar to 'Northblue.' The plant is highly self-incompatible and requires good cross-pollination. Frozen fruit quality superior to that of 'Northblue' and 'Northsky.' Fruit ripens July 11 to 17 at South Haven, Michigan (Luby, personal communication; Luby 1991).

Sunrise: G-180[G-100 ('Ivanhoe' × 'Earliblue') × 'Collins'] × Me - US 6620 [E-22 ('Earliblue' × No. 3 [North Sedgewick lowbush] × 'Earliblue')] × Me-US 24 (NH-1['Coville' × North Sedgewick lowbush] × 'Earliblue')]. 1989. 7a-5a.

Bush: Moderately vigorous, taller and easier to manage than Bluetta, upright.

Fruit: Similar to 'Bluetta' in size and color, but scar, firmness, and flavor superior to 'Duke' and 'Bluetta.'

Notes: Susceptible to stem blight in North Carolina. Resistant to red ringspot virus. Produces medium yields of good-quality, early-season fruit suitable for commercial packing and U-Pick. Ripens July 1 at South Haven, Michigan (Anonymous 1991; Draper et al. 1991).

Sunshine Blue: Open-pollinated seedling of 'Avonblue' (Pollen parent possibly 'Sharpblue' or 'Avonblue'). 10a-6b.

Bush: Wide, 3-4 ft (1 m) tall.

Fruit: Small, firm, store well, firm, and have good flavor.

Notes: Tolerates drought and a higher pH (up to 6) than all other southern highbush. Somewhat tolerant to *Phythophora*. Requires 150 chilling hours. Profuse blooming may require attention to thinning. Ripens May 10 through June 15 at Gainesville, Florida. May not be suitable for commercial production (Hartmann, personal communication).

Toro: 'Earliblue' × 'Ivanhoe.' 1987. 7a-4b.

Bush: Vigorous, upright, consistently highly productive.

Fruit: Medium size, with small, dry scars and good color and flavor.

Notes: Susceptible to stem canker in North Carolina. Begins ripening with 'Bluecrop,' but has a concentrated ripening and finishes much sooner than 'Bluecrop.' Harvest can be completed in two pickings. Buds and wood tolerate fluctuating winter temperatures very well. Recommended as a companion cultivar for 'Bluecrop.' Ripens July 15 at South Haven, Michigan (Draper, personal communication).

Chapter 8

Soil Preparation

The grower begins soil preparation immediately after choosing a suitable site. If the slope is too steep, terrace the land to allow for proper management. Drain excessively wet areas to discourage such diseases as *Phytophthora* root rot. The cost of these operations must be feasible, or an alternate site will have to be selected. Irrigation systems must be planned and underground lines installed.

Soil tests help the grower better understand the nature of his soil and how he might alter certain characteristics to benefit the plant. If the site has fairly uniform soils, mix several random samples and submit a cup of the homogenized composite for testing. Areas suspected of having somewhat different soils should be sampled separately. It is important that samples be taken from soil in the future root zone. In addition to the usual tests for nutrients, a test for soil-inhabiting insects and nematodes should also be conducted prior to planting, and the land should be fumigated as required. This is especially important on land previously planted with blueberries.

Highbush blueberries grow well in a soil pH between 4.5 and 5.2, with the optimum about 4.8 (Coville 1910). Too high a pH may cause an iron deficiency, resulting in a chlorosis, or yellowing of the leaves (Cain 1954), and may allow the ammonium form of nitrogen to be converted to the less useful nitrate form (Townsend 1967; Herath and Eaton 1968). Nitrate can increase the concentration of aluminum in the roots to toxic levels, while ammonium nitrogen will reverse the trend (Peterson et al. 1987). There is also some evidence to indicate that the nitrate form of nitrogen inhibits growth of mycorrhizae, while the ammonium form stimulates it (Korcak 1988). Mycorrhizal activity generally decreases with increasing levels of fertilization (Powell 1982). In a soil with pH lower than about 4.5, aluminum and manganese become toxic, resulting in poor bud break and some dieback.

If soil is too acidic, apply ground dolomitic limestone to raise the pH into the optimum range (Johnston 1951). Where the pH is higher than 5.2, an acidifying agent–such as very finely ground sulfur, tannic aid, or ferrous sulfate–may be used to increase acidity (Collison 1942) (Table 8-1). This will allow both iron and ammonium nitrogen to be used more effectively by the plant. Do not apply dry sulfur later than six months prior to planting, nor at a rate exceeding 1 ton per acre (1 metric tonne per hectare over a six-month period. Do not exceed 300 lbs (370 kg) per single application. Working with rabbiteye plants, Spiers and Braswell (1992) found that applications of sulfur up to 1000 pounds per acre (1120 kilograms per hectare) in the irrigation water over a single or several applications was not detrimental to plants on a high pH (6.6) soil. Though soil pH decreased, the treatment was not sufficient to lower it to desirable levels when the soil was irrigated up to seven months each year with water containing moderate amounts of sodium. Fertilizing with ammonium sulfate will also acidify the soil (Bailey and Kelley 1959).

Peat moss worked into the soil will make it more acidic over time. Wildung et al. (1990) compared the acidification efficacy of acid peat, sulfur, and a combination of acid peat and sulfur placed in the hole at planting. They found that there was no benefit to delaying planting for a year after adding the amendments. The combination of acid peat and sulfur resulted in the fastest and greatest change in

Table 8-1. Preplant recommendations to lower soil pH to 4.5.

Change pH from	Pounds of sulfur per acre (kg/ha) Soil Type			
	Sand		Loam	
7.5	950	(1045)	2800	(3080)
7.0	750	(825)	2400	(2640)
6.5	600	(660)	1900	(2090)
6.0	500	(550)	1500	(1650)
5.5	350	(385)	1000	(1100)
5.0	150	(165)	500	(550)

Note: If ferrous sulfate is substituted, multiply the amount of material applied by 6.

soil pH, the longest lasting change, the fastest plant growth rates, and the best fruit production. Plant size in the acid peat-only treatment was comparable to that in the acid peat and sulfur treatment but fruit production was less. Plants in the sulfur-only treatment showed no improvement in growth or production.

Aluminum sulfate has been used to lower the pH of soil, but it has no nutritive value. In fact, aluminum can be toxic to the plant, especially in the absence of an organic mulch, so the compound is no longer recommended (Peterson et al. 1987). Ferrous sulfate is about equivalent to aluminum sulfate in acidifying potential and it adds valuable iron to the soil. This nutrient is very important in blueberry nutrition. In addition, it leaves no toxic residue if properly used.

If blueberries are to be set in ground formerly planted to a row crop, check the soil for the presence of herbicide residues. Collect composite soil samples from the blueberry field and from land known to be herbicide-free. Plant bean seeds in each, and place them in a warm, sunny location, keeping them well watered. If bean growth is checked in soil from the blueberry plants but not in the herbicide-free soil, delay planting for at least a year.

Prepare the soil at least one season in advance, to thoroughly incorporate soil amendments, fertilizer, and organic materials and to bring weeds under control. Repeated disking will help control many perennial weeds. Certain grasses, however, such as quackgrass, are very difficult to control by this method and herbicides may have to be used.

Animal manures also can be applied about six months prior to planting and will improve soil tilth primarily by increasing organic matter. Although its efficacy for blueberries is questionable (Coville 1921; Johnston 1943), a fall application of manure at the rate of 10-20 tons per acre (22.4-44.8 metric tonnes per hectare), followed by a cover crop, will increase soil organic matter. Natural manure will give a slightly alkaline reaction. Be sure that large amounts of limestone have not been used in the stalls. Nearly all animal manures, except fresh poultry manure, are suitable and will give similar results. Poultry manure is relatively low in organic matter and can contain very high levels of nitrogen, which could damage plants. Any animal manure, particularly horse, will introduce weed seeds.

Sewage sludge may have possible future use in blueberry plantings, but its use on food crops is not recommended at this time. Seaweed has potential for use, but evidence in support of it is lacking. Also, because of its bulk and general lack of availability, it is of interest only to growers near the seashore. Use it only after leaching and on a trial basis until its suitability can be demonstrated.

Sawdust, leaf mold, and other organic material can be incorporated into the soil prior to planting (Dale et al. 1989). However, maple leaves deposit an alkaline residue upon decomposition and should not be used in great quantity. The use of organic materials in land preparation adds some nutrients to the soil and helps to conserve water and nutrients by reducing leaching. It also stimulates the activity of soil microorganisms, which release additional nutrients.

After weeds are brought under control and soil amendments thoroughly incorporated, a green manure crop such as buckwheat or millet can be planted in early summer, worked into the soil in the early fall, and followed by a winter cover crop. A complete fertilizer, such as 10-10-10-2 (2% magnesium) or 10-10-10, should be applied in sufficient quantity to produce a heavy growth and to reduce nitrogen depletion in the soil after plow-down. Base fertilizer applications on soil tests. The winter crop should be worked into the soil in early spring and the soil brought to a fine texture by thorough harrowing or rototilling. A delay in spring plowing could create problems. For example, if rye growth exceeds approximately 9-12 in (22-30 cm), there may be difficulty turning it under. Also, allowing it to become very tall (24-36 in or 60-100 cm) could retard its rate of decomposition and dry the soil considerably.

The purpose of a green manure crop is to increase soil organic matter after plow-down, to hold soil against erosion, and to suppress weed growth. Since it is not intended as a permanent crop, seeding rate can be light. Table 8-2 gives information on some green manure crops that will tolerate low pH. Generally, grasses and small grains will produce the most organic matter, while the legumes produce somewhat less organic matter (but considerably more nitrogen). The more carbonaceous (fibrous) material turned under, the more soil nitrogen that will be used by microorganisms in decomposition. This could result in depletion of nitrogen available to young blueberry plants and a subsequent check in their growth. The proportion

Table 8-2. Preplant cover crops for soils with low pH.

Crop	Seeding rate[z]	Time	Notes
Buckwheat	60	Late Spring	Reseeds when mature
Alsike clover	4	Late Spring	Nitrogen-fixing
Hairy vetch	40	Late Summer	Nitrogen-fixing
Japanese millet	20	Early Summer	Till at 20 inches
Spring oats	100	Early Spring	Frost kills
Annual ryegrass	30	Spring	Rapid establishment
Perennial ryegrass	25	Spring	High organic value
Winter rye	112	Late Summer	Till early spring
Sudan grass	80	Late Spring	Till at 24 inches

[z]Pounts per acre

Table 8-3. Approximate carbon:nitrogen ratios of organic materials.

Material	C:N Ratio
Alfalfa	12:1
Sweet clover (young)	12:1
Sweet clover (mature)	24:1
Rotted manure	20:1
Oat straw	75:1
Cornstalks	80:1
Timothy	80:1
Sawdust	300:1
Buckwheat	19:1
Peat moss	58:1
Fresh grass clippings	20:1
Dry leaves	40:1
Mature compost	15:1

of carbon to nitrogen released by material is called the C:N ratio. These ratios are given in Table 8-3 for several common soil amendments. In general, ratios greater than 20:1 indicate the need for applying preplant nitrogen fertilizers.

Research indicates (Kramer et al. 1941; Dale et al. 1989; Spiers

1980, 1983) that the use of 50/50 (V/V) peat moss/loam mix in the planting hole increases plant survival, vigor, and fruit production. The peat moss must be thoroughly soaked before incorporating, or it may dry the soil in the root zone. The use of fresh unmixed sawdust in the planting hole, as a substitute for peat moss, substantially stunts the growth of plants in the first few years (Gough et al. 1983). However, sawdust mixed with loam was not harmful. Odneal and Kaps (1990) reported that either fresh or aged pine bark could be mixed with soil and substituted for sphagnum peat in the planting hole. The use of new, absorbent starch-acrylate polymers is not recommended for blueberries, since they have a very high pH (above 8) and a high salt content (Gupton 1985). However, Austin and Bondari (1992b) reported that 'Georgiagem' plants grown in a mixture of peat moss and hydrogel produced plants with larger volume than those grown in peat moss alone. Hydrogel mixed with soil was detrimental to plant growth.

Chapter 9

Planting

If plants arrive from the nursery and cannot be set immediately, they should be unpacked and heeled in by placing the roots in a hole or trench and mounding soil around them. Soak dry root systems in a tub of water or starter fertilizer solution for several hours prior to heeling-in. If the ground is frozen, put the plants in a cool, well-protected area, such as an unheated basement or garage, and cover them entirely with damp peat moss or sawdust.

Blueberries can be planted in the fall in areas that experience heavy snowfall or winters in which temperatures do not fall below 0°F (–18°C). Planting at this time makes weed control easier and allows a few months for new root development and growth before spring bud break. However, fall planting in areas where winters are harsh can result in severe heaving and winter damage. Planting in early spring, as soon as the ground can be worked, will avoid this. Those plants set at the earliest possible date often make the best growth. Dormant plants withstand transplanting better than those in leaf because the root system has no leaf area to supply with water. Newly set plants in leaf must be watered frequently to prevent wilting, which of course increases grower effort. These types of plants are also quite susceptible to frost damage. Always keep the plants' root systems moist while in the field prior to planting, to reduce root and leaf desiccation. Planting on a cloudy afternoon is best, but is not always possible, especially with larger plantations.

It makes little difference whether rows run east to west or north to south, so long as plants are protected from strong winds. If plants are to be managed with clean cultivation, they should be oriented across steep slopes to reduce soil erosion. However, planting in hedgerows across slopes can interfere with air drainage and increase frost damage. If fields are long and narrow, many growers prefer rows running lengthwise, to reduce the number of turnarounds for equipment. Allowances must be made at each end of the field for turnaround (headland). To figure the amount needed, measure the distance across the field and divide by the planting distance within the row. For example, if bushes are spaced 5 ft (1.5 m) apart and the field is 74 ft (22 m) long, then 14 plants will fit in a row, with 4 ft (1.2 m) left over (74 ÷ 5 = 14, with a remainder of 4). This is not enough for turning, so rows are subtracted and their distances added to the remainder until enough room remains. If five rows are removed (25 ft or 7.5 m) and the distance added to the remainder of 4 ft (1.2 m), the resulting 29 ft (8.7 m), when divided by two to allow the same amount on each end, yields an ample headland of 14 ft 6 in (4.3 m).

When replanting former blueberry plantations, do not plant in the same row from which old plants were removed. Set new plants between former rows, or, if this is not feasible, wait at least three years after the removal of old plants before setting new ones. These practices will reduce the incidence of carry-over pests infesting new plantings.

For greatest yields, avoid planting large blocks of a single cultivar, since this may not allow for adequate cross-pollination. Also, because bees tend to fly along rows instead of across them, omit planting a bush every 25 bushes, to leave a "fly-through" for the bees.

Young plants are usually set 3-6 ft (0.9-1.8 m) apart in rows spaced 8-10 ft (2.4-3.0 m) apart. Exact spacing depends upon cultivar, the type of equipment used for management and harvesting, and upon the natural fertility of the soil. Sandy soils with low fertility allow closer spacings. Because highly fertile soils produce larger plants, in-row distances should be increased. The distance between rows can be estimated by adding about 6 ft (1.8 m) to the driving width of the equipment. Plants should be spaced far enough apart so that they do not touch and interlace within the row. This hedgerow effect reduces air circulation, light penetration, and flower bud formation. Refer to Appendix 2 for the number of plants required to plant an acre at various spacings.

Plants are usually set at the same depth as they were in the nursery. Planting them more than 2 in (5 cm) deeper can suffocate the root system and kill the plant (Gough 1983b). This is especially true in heavy soils or after heavy, soil-compacting rains.

Dig planting holes large enough so that no root crowding occurs. Usually, no root pruning is required. However, broken large roots on older plants should be cut above the break, and excessively long roots should be trimmed. Never wrap long roots around the root ball to fit into the hole. If plants are purchased in containers, gently break up the root ball before planting. Reject any plants that appear pot-bound, since these seldom fully recover in the field (Clayton-Greene, personal communication). If plants are balled and burlapped (B&B) with real burlap, untie and loosen the top after placing the plant into the hole. It is not necessary to remove the burlap completely. Plants balled in the newer plastic weave materials that resemble burlap should be removed from their wrapping prior to planting.

Planting holes can be made with a spade or with a tractor drill used for digging post holes. However, the latter may compact and glaze the walls of the hole, particularly in heavy soils. These holes must then be scarified with a hoe or spade to allow better drainage and root penetration. In lighter soils, plants can be set in a plow furrow.

After setting the plant, fill the hole three-quarters full with soil or peat/soil mix and flood it. When the water has soaked in, fill the rest of the hole, tamp firmly, and water again with plain water or a weak starter solution (Gough, unpublished data). Never place dry, standard granular fertilizer or manure into the planting hole, since they can easily injure the young root system. Australian growers have had good results by mixing two handfuls of peat and 0.7-1.0 oz (2-3 g) each of six- to nine-month and three- to four-month osmocote into the planting hole (Clayton-Greene 1989).

Plants in low areas can be set on mounds or raised beds within rows to reduce the potential for spring flooding of their root systems (Doehlert 1937). After planting, shoots of bare-rooted plants should be cut back by half or to two or three buds to better balance the top with the root system (Meador et al. 1983). Also, any dead or injured branches should be removed.

Chapter 10

Soil Management

The two most important practices in every new planting–proper irrigation and good weed control–are often approached halfheartedly by the grower (Gough 1982).

Highbush blueberries require a constant moisture supply. A mature planting should receive about 1-3 in (2.5-5 cm) of water per week during the entire growing season, or at least through the harvest period. This requires about 600-1200 gal per 1000 sq ft (2280-4560 l per 93 sq m). Traditionally, selection of a site with a water table 18-24 in (45-60 cm) below the soil surface has met demands for water. For extra insurance, supplementary irrigation is strongly recommended in most growing regions and is considered absolutely mandatory in the Pacific Northwest area of the United States (Martin, personal communication).

Commonly, two types of supplemental irrigation systems are used. The sprinkler system provides sufficient water for proper growth and can be used to provide protection from frost. Use caution, however, during fruit ripening. Water applied to a ripe berry's surface can lead to fruit cracking. Further, water pressure from sprinklers can cause some mechanical damage and knock ripe fruit from plants.

Trickle irrigation, also known as microirrigation and drip irrigation, has become popular in recent years. This system supplies water only to the plant and not to areas void of roots. Because water is supplied at ground level, evaporation is reduced and water conserved. Fruit remains dry. This allows watering through harvest and reduces the incidence of cracking and mechanical damage. Because of low water requirements and pressures, operating expenses are lower than those for an overhead system. However, trickle irrigation provides no means of frost protection. It could also lead to inconsistent productivity unless properly engineered. Gough (1984a,b) and Abbott and Gough (1986) reported that potted plants receiving water on one side only grew and fruited on that side, while growth was severely stunted on the dry side. Patten et al. (1989) reported that blueberry plants grew best in the field when watered on all sides, compared with plants receiving water beneath only a small segment of their dripline. In Australia, Shelton and Freeman (1989) found that during dry periods the side of the plant opposite the emitter showed drought stress. The grower should take care to place water emitters so that the entire root system is thoroughly irrigated.

Weeds compete with blueberry plants for water and nutrients; shade young plants, decreasing their ability to photosynthesize; and harbor insects and diseases. Further, they decrease air circulation around the bushes and can interfere with pesticide applications. Weeds must be controlled for best production. There are several methods for doing this, each with its own advantages and disadvantages.

SOD

The plantation may be kept in sod. This retards erosion and makes operation of equipment easier during very wet or dry periods. Further, sod is cleaner and more pleasant than soil to walk on. The sod should be kept mown to maintain a neat appearance and to keep heavy growth from competing with the bushes. Plantings under this system are often colder in spring and more prone to frost damage, since the sod insulates the soil and retards loss of ground heat to the atmosphere. Further, control of such diseases as mummyberry can be more difficult in sod, and crops can be substantially decreased in dry years on sandy soils without supplemental irrigation (Johnston 1937).

CLEAN CULTIVATION

Shallow cultivation (about 2 in or 5 cm) practiced frequently enough during the growing season will help aerate the soil, keep heavy weed growth down, and generally stimulate plant growth (Coville 1921). It should begin in early spring and continue as necessary until berries begin to ripen. Cultivation during ripening can knock berries off the bush. Cultivation after late summer is not recommended in cold areas since it can stimulate late growth, which often fails to harden sufficiently (Bailey et al. 1939). Clean cultivation can lead to soil erosion on slopes. Freshly disked soil (1) remains about 2°F (1°C) colder than firm, bare ground, (2) does not reduce the potential for frost damage, (3) can make walking unpleasant, and (4) can require greater amounts of irrigation. However, it can also make control of certain diseases like mummyberry simpler, because it eliminates debris beneath the plant. Under constant cultivation, soil nitrate levels increase and the soil pH becomes more alkaline, while ammonium nitrogen levels decrease. Blueberries make their best growth where ammonium nitrogen levels are high (Cain 1952; Peterson et al. 1988).

COVER CROPS

Practicing clean cultivation during the growing season and sowing a cover crop after harvest helps harden plants for winter; it decreases winter erosion; traps snow and helps insulate the soil from extremely low temperatures; adds organic matter; reduces soil caking; and promotes better water penetration. A legume crop such as vetch or clover can also add nitrogen to the soil. Spring oats make a good non-legume crop. Winter rye should not be used because it will overwinter and become too difficult to subdue the following spring. Cover crops can harbor diseases and insects and deplete nutrients and water. Do not use them in drought areas without irrigation.

MULCHING

This is perhaps the wisest and most widely used soil management practice on smaller plantations, and it is highly recommended. Among other things, organic mulches can reduce frost damage by slowing spring bud development and can delay leaf drop in the fall, allowing plants more time to store carbohydrates and harden-off (Patten et al. 1989). Some are also more effective than sulfur in acidifying the soil (Cummings et al. 1981) and helping maintain lower solubilities of toxic aluminum and manganese (Korcak 1988). Apply at least 6 in (15 cm) of organic mulching material. Dried grass clippings, peat moss, buckwheat hulls, shredded leaves, straw, wood chips, and sawdust are suitable. Leguminous hay has some-times proven injurious to the blueberry plant by stimulating late-sea-son growth. Straw mulch can also increase the levels of nitrate nitrogen in the soil and decrease the preferred ammonium levels. Because of potential problems with availability and expense, the first four are suitable for small plantings only. Grass clippings from areas previously treated with herbicides should not be used. Also, to avoid fermentation and overheating, do not use fresh grass clip-pings. Peat moss and buckwheat hulls are relatively expensive and the former may crust, reducing moisture penetration to the plant roots. Any organic mulch increases the chances of mice injuring the crown of the plant in the winter and can deplete soil nitrogen during decomposition. Of the last three materials mentioned, sawdust and wood chips give very good results, though sawdust is more apt to erode than chips. Lareau (1989) reported that sawdust mulch was superior to any other factor in growth and productivity of blueberry plants on mineral soils. Peterson et al. (1987) found that without the use of a sawdust mulch, aluminum sulfate used to acidify the soil restricted blueberry root growth.

Under sawdust, the soil acidity remains nearly constant, and the ammonium form of nitrogen increases. Soil temperatures are higher in winter and lower in summer than they are without the mulch, and daily temperature fluctuations are decreased. Soil moisture content is greater under sawdust than under clean cultivation (Shutak and Christopher 1952; Lareau 1989), and sawdust reduces soil compac-tion and increases aeration (Lareau 1989). Plant size and crop yields substantially increased when sawdust mulch was used on highbush (Christopher and Shutak 1947), southern highbush (Clark and Moore 1991), and lowbush blueberries (Sanderson and Cutcliffe 1991). Also, the reduction in maintenance is enough to at least partially compensate for the cost of the mulch. The type of sawdust used may affect the cost. Softwood sawdust is usually coarser in texture and decomposes more slowly than hardwood sawdust. It also does not pack as readily as hardwood. Depending upon the material, additional quantities, usually about 1 in (2.5 cm), should be added each year to compensate for that lost in decomposition (Boller 1956).

When sawdust is used, spread it to a depth of 6 in (15 cm) over the entire planting, or at least in a 2-ft (60-cm) radius circle beneath individual bushes, slanting it toward the plant to decrease water runoff away from the root zone. For small plantings, use about 5 bu (176 l) of sawdust per plant. This is equivalent to about ten 5-gal (19-l) buckets. Twenty to thirty cords per acre (185-278 cu m per ha) are required to mulch a 3-4-ft (1-1.3-m) ring around each bush; 35-50 cords per acre (323-463 cu m per ha) are needed to mulch a strip 3-4 ft (1-1.3 m) wide along each row; and 100 cords per acre (925 cu m per ha) are required for complete coverage (Gough et al. 1983). It is usually considered best not to use fresh sawdust though Clark and Moore (1991) reported no problems with it. Sawdust that has been piled and allowed to decay for a few years should be turned, aerated, and leached for several months before use.

Because soil microorganisms under mulching conditions use some of the available soil nitrogen to break down the mulch, nitrogen deficiency can develop. Apply additional nitrogen to correct it. This can be done by doubling the normal amount of fertilizer, or by adding about 1.5 lb (.7 kg) of 21-0-0 per 100 lb (45 kg) of mulch.

Organic mulches increase the yield of blueberries, but may not be desirable because of difficulty in their procurement and handling, their potential expense, and in problems that could arise from their use (such as rodent damage and fire hazard). The use of an alternative mulch, such as plastic, has gained some interest in recent years. Black polyethylene mulch placed on the soil surface, or over a sawdust mulch, and around newly set plants effectively controls weeds, but makes fertilizing difficult. Also, temperatures beneath the plastic can rise 125% above those of the ambient air, resulting in severe damage to the root system. Plant death is more likely with black plastic than with organic mulches if supplemental irrigation is not provided. (Gough, unpublished data). In Gough's trials, yields were less than those for plants mulched with sawdust alone. When the plastic mulch was placed directly on bare soil and then covered with sawdust, blueberry plants produced two root systems—one above and one below the mulch, in four years' time. When conditions were conducive to growth, these plants grew faster than plants mulched in a different manner. However, when the upper root system was killed by severe drought, the plants suffered substantial dieback. Use of black polyethylene as a short term mulch has proven effective if enough fertilizer is incorporated into the soil and if provisions for trickle irrigation lines are made before mulch is applied (Mainland and Lilly 1984). The effective life of the plastic exposed to sunlight is about two seasons. This is long enough to allow plants to become established. Australian researchers have reported that black polyethylene was superior to sawdust and pine bark for establishing plants, but eventually should be replaced by either of those two after the establishment period (Clayton-Greene 1988).

In recent years, researchers have examined the uses of living mulches for blueberry production. These crops are grown as cover crops between rows and mowed, the clippings placed beneath the blueberry plants as a mulch. The species of mulch used depends upon geographical location and time of year it will be grown. Georgia researchers have found pearl millet to be successful in summer and rye or annual ryegrass in winter. Crimson clover was also found suitable. There should be a buffer zone of 8-20 in (0.2-0.5 m) between the cover crop and the blueberry plants to decrease summer competition (Patten et al. 1990; Pavlis 1991).

HERBICIDES

In most cases, the use of mulches and mechanical hoeing will give satisfactory control of weeds. However, both can be labor-intensive, and the grower may have to resort to less expensive chemical weed control. Regulations governing the use of herbicides vary from time to time and place to place. Therefore, always follow instructions on the label and the most recent recommendations from your local Cooperative Extension service or state university. Be sure your sprayer is in good condition and properly calibrated before spraying begins. Remember, herbicides kill plants, as their name implies. Misguided sprays can kill blueberry plants as well as weeds.

The following are some herbicides that have been used effectively to control weeds in blueberry plantings (Meade 1989, 1990, 1991).

Chloropham (isopropyl m-chlorocarbanilate)
Product Name: Chloro IPC
Use: Preemergent applications to control annual grasses and broadleaf weeds. Ragweed and members of the Composite family are tolerant. Often used in conjunction with other preemergent herbicides.
Mode of Action: It is absorbed through emerging coleoptiles and roots and translocated throughout the plant. It disrupts cell division, interferes with respiration and photosynthesis, and inhibits protein synthesis.
Persistence: This is greatly affected by microbial activity and soil organic content and moisture, but half-life is about 65 days at 59°F (15°C) and 30 days at 84°F (29°C).

Dichlobenil (2,6-dichlorobenzonitrile)
Product Name: Casoron, Norosac, or Dyclomec
Use: Preemergent applications during cool weather to control broad spectrum of weeds. Do not apply to frozen ground.
Mode of Action: It is absorbed from the soil by the roots, or through the leaves during sublimation. It disrupts meristematic and phloem tissue, resulting in collapse of the stem.
Persistence: Half-life varies from 1.5-12 months, depending upon soil type.

Diuron (3-(3,4-dichlorophenyl)-1,1 dimethylurea)
Product Name: Karmex
Use: This controls certain weeds when applied as a preemergent or early postemergent spray. It will not adequately control established weeds. Use in established fields at least one year old.
Mode of Action: It is absorbed through the roots, translocated in the xylem, and strongly inhibits photosynthesis.
Persistence: About one season.

Fluazifop-butyl ((+)-butyl 2-[4-[[5-(trifluoromethyl)-2-pyridinyl] oxy]phenoxy]propanoate)
Product Name: Fusilade
Use: It is a highly selective postemergent annual and perennial grass killer, but it will not control broadleaf weeds. Should be used on non-bearing planting only.
Mode of Action: It is absorbed through the leaf surface and translocated through phloem and xylem into rhizomes and stolons.
Persistence: Up to 60 days.

Glyphosate (N-phosphonomethyl glycine)
Product Name: Roundup
Use: A very broad-spectrum, non-selective herbicide that is very effective on deep-rooted perennials, grasses, and broadleaved weeds. It should be applied as a postemergent spray to weed foliage. Effectiveness can be increased by adding a small amount of spray-grade ammonium sulfate to each tankful. Do not allow contact with green blueberry tissue.
Mode of Action: It is absorbed through the foliage and is translocated throughout the plant.
Persistence: Relatively non-persistent.

Hexazinone (3-cyclohexyl-6-(dimethylamino)-1-methyl-1,3,5-triazine-2,4(1H,3H)-dione)
Product Name: Velpar
Use: A broad-spectrum, non-selective pre- or postemergent herbicide that controls annual, biennial, and perennial grasses and certain broadleaf weeds. Low soil organic matter can result in Velpar damaging the blueberry plant.
Mode of Action: It is absorbed through both roots and foliage, moves in the xylem, and appears to inhibit photosynthesis.
Persistence: Persistent up to six months under some field conditions.

Napropamide (2-(a-naphthoxy)-N,N-diethylpropionamide)
Product Name: Devrinol
Use: It is a preemergent herbicide that controls most annual grasses and many broadleaf weeds. It should be applied just before a rain or irrigation and incorporated into the top 1-2 in (2.5-5 cm) of soil.

Mode of Action: It is absorbed by emerging roots and translocated to the ariel portions of the weed. It inhibits root growth.

Persistence: It can remain in the soil more than 12 weeks if incorporated after application.

Norflurozon (4-chloro-5-(methylamino)-2-(a,a,a-trifluro-m-tolyl)-3)2H)-pyridazinone)

*Product Name:*Solicam

Use: It is a preemergent herbicide for grasses, rushes, sedges, and broadleaf weeds. Follow application with a broadleaf herbicide. Use different herbicides in alternating years.

Mode of Action: It is absorbed by the roots and translocated to actively growing portions of the plant, where it inhibits carotenoid synthesis. Without these to filter sunlight, chlorophyll is destroyed and photosynthesis is inhibited.

Persistence: It remains in soil up to 200 days, depending upon soil structure.

Oryzalin (3,5-dinitro-N4,N4-dipropylsulfanilamide)

Product Name: Surflan

Use: It is a selective preemergent herbicide for controlling many annual grasses and broadleaf weeds. It should be incorporated into the soil by watering or shallow disking.

Mode of Action: Affects seed germination.

Persistence: It is not persistent, and it rapidly biodegrades.

Paraquat (1,1'-dimethyl-4-4'-bipyridinium ion)

Product Name: Paraquat, Gramoxone

Use: It is a general, non-selective postemergent contact herbicide.

Mode of Action: It is absorbed rapidly through the foliage and translocated through the xylem.

Persistence: It is rapidly deactivated in the soil.

Simizine (2-chloro-4,6-bis(ethylamino)-s-triazine)

Product Name: Princep, Simazine, Primatol

Use: It is a very effective preemergent herbicide for control of broadleaf weeds and grasses. For best results, apply just before a rain or irrigate after application. Do not apply after fruit set.

Mode of Action: It is absorbed by the roots, translocated by the

xylem, and accumulates in the leaves and apical meristems, where it inhibits photosynthesis.
Persistence: Relatively short.

Terbacil (3-tert-butyl-5-chloro-6-methyluracil)
Product Name: Sinbar
Use: It should be applied to the soil surface while blueberries are still dormant to control a broad spectrum of weeds. Apply to established plantings only, prior to bud break. Do not allow contact with blueberry plants.
Mode of Action: It is most readily absorbed through the root system and moves into ariel portions, where it inhibits photosynthesis.
Persistence: May persist nearly a year in some soils.

FERTILIZERS

Because growing plants remove some nutrients from the soil, these nutrients must be replenished with fertilizer to keep the plants vigorous. Fertilizer should always be applied beneath the dripline in a broad band. Since it does not move horizontally in the soil to any appreciable extent, proper placement is important. Raking or disking it in will aid in its becoming available more rapidly. There is little lateral transport of water and nutrients from one side of the bush to the other, and fertilizer applied to one side of the plant will primarily fertilize that side only (Gough 1984a,b). Be sure to apply fertilizer completely around the bush or at least on both sides of the row. Browning of the leaf tips and margins can indicate too much fertilizer; placement of fertilizer is too close to the plant crown; uneven distribution of the fertilizer; application of fertilizer in dry weather; or the use of fertilizer with a very high salt index (Table 10-1). Though they may provide useful nutrients for plant growth, high-index fertilizers have a greater potential for burning foliage and roots and must be used with extreme caution. Cummings (1989) found that growers had better success using higher analyses fertilizers because of their lower salt index. Effects of potentially damaging fertilizer applications may be lessened by heavy watering, which reduces the concentration of fertilizer in the soil by leaching.

Blueberry has a relatively low nutrient requirement and is sensi-

Table 10-1. Relative salt effects of fertilizer on soil solutions.

Material	Salt Index[z]
Ammonium nitrate	104.7
Ammonium sulfate	69.0
Diammonium phosphate	29.9
Monoammonium phosphate	34.2
Potassium nitrate	73.6
Potassium sulfate	46.1
Sodium nitrate	100.0
Sulfate of potash-magnesia	43.2
Superphosphate (20%)	7.8
Superphosphate (45%)	10.1
Urea	75.4

[z]The higher the index, the greater the osmotic pressure and the potential for injury.
Source: Rader et al. 1943.

tive to overfertilization. The amount of fertilizer required for best plant growth depends upon crop growth, soil structure, leaching, erosion, and soil management system. Soil samples should be taken every few years to determine nutrient reserves and to better enable the grower to adjust the fertilizing schedule. Collect soil samples in the fall from the root zone near the dripline of several plants distributed randomly over the entire area. Problem areas should be sampled separately. Commonly, your Cooperative Extension Service will supply test results for soil pH, phosphorus, potassium, soil organic matter, and nitrogen, although soil test results are not particularly useful for determining nitrogen needs. Observation of shoot growth and fruit production, together with foliar analysis, should be used for this. The grower should also ask that magnesium and calcium tests be made. A magnesium to calcium ratio of 1:10 and a potassium to calcium ratio of 1:5 are usually considered about right for blueberry soil.

Foliar analysis is more useful than soil analysis, since it determines amounts of nutrients actually being taken in by the plant. Leaf sampling should be conducted during the four-week period ending with the first week in which the first third of the crop is harvested.

Leaves for analysis should be the youngest, fullest-sized leaves harvested from nodes 4, 5, and 6 from the top of fruiting shoots. Select random samples from each of ten plants for each sample to be tested. Wash in clear water, rinse, and dry before sending to the laboratory for analysis. Do not allow them to lie in water, since some nutrients may be leached from the tissue.

The appearance of apparent mineral deficiency symptoms may be the result of an actual deficiency or may result from several other conditions or combinations of conditions. These include: poor soil moisture distribution; poor drainage and subsequent restriction of the root system; insects; disease; fertilizer burn; weeds; and compaction of the soil. All of these conditions can weaken the root system. Other problem conditions include insufficient quantities of available ammonium nitrogen; periods of cool weather during the growing season; injury from pesticides; and erosion of the topsoil. Be sure that these conditions do not exist before making additional fertilizer applications. Table 10-2 will help you determine the nutrient needs of your plants, based on foliar nutrient levels.

The individual grower will need to establish his own rate of fertilization. This can be done by watching for signs of good plant

Table 10-2. Suggested foliar nutrient levels for blueberry.

Element	Deficiency below	Standard Minimum	Standard Maximum	Excess above
Nitrogen	1.70%	1.80%	2.10%	2.50%
Phosphorus	0.10%	0.12%	0.40%	0.80%
Potassium	0.30%	0.35%	0.65%	0.95%
Calcium	0.13%	0.40%	0.80%	1.00%
Magnesium	0.08%	0.12%	0.25%	0.45%
Sulfur	0.10%	0.13%	0.20%	--------
Manganese	23 ppm	50 ppm	350 ppm	450 ppm
Iron	60 ppm	60 ppm	200 ppm	400 ppm
Zinc	8 ppm	8 ppm	30 ppm	80 ppm
Copper	5 ppm	5 ppm	20 ppm	100 ppm
Boron	20 ppm	30 ppm	70 ppm	200 ppm

[z]Leaf nitrogen levels above this may cause a reduction in yield.
Source: Doughty et al. 1981; Ballinger et al. 1958.

vigor and for certain nutrient deficiency symptoms. In general, mature plants should produce several whips near ground level and laterals 12-18 in (30-45 cm) long. The most productive shoots have 15-20 leaves (Gough and Shutak 1978). Poor vigor and small, pale green leaves that color and drop early in the fall often indicate insufficient fertilization. Although about a dozen nutrients have been reported deficient in blueberry plants, many of these deficiencies have been induced in sand culture under greenhouse conditions. Because the majority of these have never been observed under field conditions, it does the grower little good to study them. Those most commonly encountered in the field are nitrogen, magnesium, iron, and, to a lesser degree, potassium and boron.

Nitrogen

A deficiency of nitrogen, the nutrient most often lacking under field conditions, results in a uniform paleness or yellowing of the entire leaf surface (Hull 1967). Leaves then redden and die. In less severe cases, leaves will redden and drop early in the fall. Since nitrogen is moved from older to younger tissues, older leaves often show symptoms first. Young shoots arising from the base of the plant appear pink at first, then turn pale green. The entire bush is stunted. Requirements vary according to geographic location, soil type, plant age and productivity, and soil management practice. Organic mulches require additional nitrogen applications. A common recommendation for mature highbush blueberry plantings on upland, mineral soils is about 40 lb of actual nitrogen per acre (45 kg per ha) per year. This amount must always be "fine-tuned" by visual examination of the plants under specific site conditions, since excessive rates can reduce fruit production, delay ripening, damage plant growth, and increase winter damage.

The form of nitrogen fertilizer is important to blueberry growth. Plants receiving nitrate as the sole nitrogen source make poor growth and can show both nitrogen and iron deficiencies, since nitrates will gradually raise the soil pH. They are easily leached, absorbed poorly by blueberry plants, and can become toxic in high concentrations. The use of ammonium nitrate as the sole nitrogen source reduces yields and results in smaller berry size (Hull 1967). Adding nitrogen in the ammonium form can increase plant growth and remove deficiency symptoms (Cain 1952; Takamizo and Sugi-yama 1991), in part because that form lowers soil pH. This is the form of nitrogen fertilizer preferred for blueberries and is usually supplied as ammonium sulfate or urea. When applied in tempera-tures above 80°F, about a third of the urea can be lost through volatilization if not incorporated quickly into the soil. Retamales et al. (1989) found that urea applied to the soil before spring bud break appeared in the leaves within two weeks, and reached its greatest concentration within three weeks after application. By season's end, the blueberry plant had recovered about 32% of the nitrogen ap-plied, storing it mostly in the leaves and young shoots. Less than 15% remained in the soil. A foliar spray of urea applied at the rate of 4-6 lb of actual nitrogen per acre (4.5-6.75 kg per ha) in 150-200 gal (570-760 l) of water is useful for quick relief of nitrogen deficiency, though Hancock et al. (1984) found that foliar feeding of nitrogen was not as efficient as soil applications to mature blueberries. Be sure to allow the urea/water mixture to return to ambient tempera-ture before spraying. Pavlis (1991b) has recommended applications of high-nitrogen foliar fertilizers in years when a large crop is ex-pected and/or when drought conditions make dissolution of dry fertilizer suspect.

Korcak (1988) suggests that excessive use of ammonium fertiliz-ers could acidify the soil to a point where increased aluminum and manganese availability would mask any benefit of the fertilizer.

Phosphorus

Leaves on plants deficient in this element appear dull and slightly purple. Where soils have a low reserve, applying phosphorus at the rate of about 30 lb of actual phosphorus per acre (34 kg per ha) can increase production.

Potassium

Potassium deficiency causes the margins of leaves to turn red, then become necrotic. Necrotic spots appear in older leaves, while interveinal chlorosis can occur on younger growth. On soils with low levels of this nutrient, additions of about 40 lb of actual potassium per acre (45 kg per ha) can increase production. In some cases, potassium chloride has killed plants, resulted in greater winter damage, and reduced fruit size (Slate and Collison 1942; Townsend 1973). Eck (1983) found plants responded well to applications of potassium in New Jersey.

Magnesium

Where this nutrient is deficient, leaf margins become chlorotic, while the midrib remains green. The chlorosis advances into a reddening, followed by necrosis. This appears first on older leaves. The symptoms usually appear during berry ripening. High potassium levels in the soil can aggravate magnesium deficiency. Applications of magnesium sulfate (Epsom Salts) or sulfate of potash-magnesia to the soil every 2-3 years, at the rate of about 200 lb per acre (224 kg per ha), can eliminate symptoms (Eck 1964). Where this deficiency is a continuous problem, adding 2% magnesium oxide to the regular fertilizer can increase production (Gough et al. 1983).

Boron

Boron deficiency has been periodically reported. Symptoms appear rapidly. First, shoot tips appear bluish, followed by chlorotic splotching of upper leaves. Leaves may become misshapen. Foliar application of boron is recommended. The use of Borax at 3 lb per 100 gal (0.36 kg per 100 l) water or Solubor at 1.5 lb per 100 gal (0.2 kg per 100 l) have been effective.

Iron

Iron-deficient plants display interveinal chlorosis, which appears first on the youngest leaves. In severe cases, the leaves may turn completely lemon-yellow or reddish brown. Basal leaves will be stunted, and new shoots will be lemon-yellow in color. High soil phosphorus may aggravate iron deficiency (Arnold and Thompson 1982). Because the availability of iron to the plant depends upon low soil pH, acidity should be checked first and corrected if necessary. Rosen et al. (1990) reported that at a soil pH of 4.5, the iron content in blueberry roots was up to 100 times higher than in the shoot. The presence of ammonium nitrogen appears essential for proper iron metabolism, and the use of ammonium nitrogen fertilizers can reduce the number of iron deficiencies. Short-term treatments include soil applications of ferrous sulfate (34% iron) at 15 lb per acre (17 kg per ha) or iron chelate at about 20 lb per acre (22 kg per ha) (Doughty et al. 1981). Excessive use of iron chelates under high pH conditions (6.5) can restrict plant growth (Korcak 1988).

Fertilizer Recommendations

Although fertilizer recommendations vary according to soil and location, most experts agree on a 1-1-1 ratio mixed fertilizer, such as 10-10-10 or 15-15-15. At least half of the nitrogen should be in organic form, which becomes available to the plants over a period of time. The fertilizer should also contain about 2% magnesium oxide, represented by a fourth number, such as 10-10-10-2.

Of all the essential elements, nitrogen gives the greatest response in blueberry plantings, while additional phosphorus and potassium often result in little improvement in growth. Therefore, some sources recommend applying only nitrogen, preferably in the ammonium form. Bushes can bear substantially heavier crop loads under higher rates of nitrogen fertilization. However, excessive rates can reduce yields and increase winter damage. Hanson and Retamales (1992) reported no difference in berry yields among plants treated with urea or two controlled-release fertilizers.

Blueberry fertilizers should have a decidedly acid reaction (Table 10-3). Avoid compounds such as nitrate of soda, cyanamid, calcium nitrate, bone meal, and wood ashes because of their alkaline residues. Most other fertilizers leave acid residues. Ammonium sulfate is often used when the soil pH is above 5.2, and urea is used when it is below 4.6. Some agencies recommend not using a fertilizer containing muriate of potash (potassium chloride), since the chlorine can be injurious under some conditions (Slate and Collison 1942).

Well-rotted barnyard manure (8-12 tons per acre of 18,000-27,000 kg per ha) or poultry manure (5-6 tons per acre or 11,000-13,500 kg per ha) can also be used, but should be applied only in late fall or very early spring. Manures will leave a slightly

Table 10-3. Effect of fertilizers on soil reaction.

Material	Pounds of ground lime needed to counteract acidity per 100 pounds material
Acid-forming	
Ammonium nitrate	60.0
Monoammonium phosphate	59.0
Diammonium phosphate	70.0
Ammonium sulfate	110.0
Phosphoric acid, liquid	110.0
Urea	84.0
Blood, dried	22.0
Castor pumice	6.0
Cottonseed meal	10.0
Fish scrap	5.0
Alkaline-forming	
Calcium nitrate	20.0
Potassium nitrate	23.0
Sodium nitrate	29.0
Bone meal	20.0
Calcium cyanamide	63.0
Cocoa shell meal	2.0
Manures, dried	15.0
Peanut hull meal	6.0
Rock phosphate	10.0
Tobacco stems	20.0
Neutral	
Calcium sulfate (Gypsum)	
Potassium chloride	
Potassium sulfate	
Superphosphate	
Sulfate of potash-magnesia	
Monopotassium phosphate	

Source: Pierre 1933.

alkaline residue in the soil and their continued use should be monitored carefully.

Newly set plants can be fertilized after the second flush of growth has begun. This often occurs a few weeks after planting. Earlier fertilization can cause reddening of the leaves and a delay in the start of new growth. Apply the equivalent of about 1 oz (28 g) of a

10-10-10 fertilizer per bush, spread around the plant at least 4 in (10 cm), but not more than 12 in (30 cm) from the crown. A second application can be made in mid-June to mid-July. Where winters are not too severe, make a third application in late November if plants show low vigor. The last two applications can be made at the same rate as the first. Always consider the conditions in your area before making the November application; it is not appropriate for all areas.

Increase the amount of fertilizer applied each year until mature bushes (after 6 years in the field) are receiving about 1 lb (0.45 kg) per plant, one-half applied at the beginning of bloom, the other 5-6 weeks later. In areas with low-organic, sandy soils and heavy rainfall, lighter, more frequent fertilization is beneficial.

If mature bushes show low vigor, an additional one-half lb (0.22 kg) per bush applied in late fall will build reserves in the plant and allow a quick growth spurt in spring. Avoid this practice on vigorous bushes, when ground is frozen, and in cold areas where winters are severe. Cummings (1989) reported that multiple fertilizer applications of up to four per year clearly increased yields in North Carolina. He attributed this increase to the additional phosphorous made available. Hanson and Retamales (1992) reported that split applications of urea (half applied at bud break and half at petal fall) resulted in 10% higher yield than the same amount in a single application at bud break under Michigan conditions.

For larger plantings, broadcast fertilizer at the rate of about 40-65 lb of actual nitrogen per acre (45-73 kg per ha) total per year. Florida growers apply 10 lb of (4.5 kg) of nitrogen in each of eight applications per year because of thin sandy soils and rapid leaching.

Chapter 11

Pruning

The object of pruning the blueberry plantation is to consistently produce good yields of high-quality fruit. Regular pruning helps control plant growth and allows more sunlight to penetrate the plant canopy, increasing photosynthesis and flower bud formation. Spray coverage and air circulation within the plant's canopy are improved, decreasing the incidence of disease and improving fruit quality. Pruning removes some flower buds, decreasing the tendency to overbear and potentially increasing fruit size, particularly during drought conditions. Because the blueberry produces several times the number of flowers required for a good crop, removal of some buds during pruning can increase the quality of fruit set on the remaining buds and also concentrate fruit ripening (Mainland 1989b). It also reduces stress on plants during dry conditions. Further, a heavy crop load can bend canes to the ground, resulting in picking difficulties and dirty berries. This is most likely to occur on cultivars such as 'Harrison' and 'Bluecrop' that have a spreading growth habit. Lastly, the process removes dead, injured, and unproductive old wood, making harvesting easier.

Although it appears to increase total plant growth by increasing shoot length and leaf size, as well as stimulating new cane development, pruning actually reduces the total vegetative growth. Total root growth is reduced, and heavy pruning over several years can be quite detrimental to the root system. Though the response to pruning is generally distributed throughout the plant, it is most evident nearest the cut.

Excessive pruning of young plants delays the onset of flower bud formation, while it increases shoot vigor. Where nitrogen is deficient and plants are stunted, pruning increases the nitrogen supply available to remaining growing points and allows for more vigorous shoot growth and flower bud formation. In some fruit species, pruning increases the percentage of fruit set.

Pruning varies in three ways: season, severity, and distribution. Because it is more an art than a science, opinions are many and varied. No two people will prune a bush the same way, nor would the same person prune a bush the same way twice if it were possible.

SEASON

It makes no difference to the plant when or if it is pruned, but, because the act disturbs internal water and nutrient balance, the response will vary slightly over the year. Blueberries should be pruned annually during late winter and early spring in northern areas, though dead, injured, and diseased branches should be removed promptly throughout the season. Some growers prefer to prune in the fall, since leaves on weak wood turn color first and the canes are easily identified and removed. The actual time of pruning can affect the blooming period (Gough 1983c). Plants not pruned will bloom earliest, while plants pruned in September (immediately after harvest) of the year before bloomed up to a week later. Those pruned in February, the beginning of normal pruning in the northeastern United States, bloomed early, shortly after those that were not pruned. Early pruning, by delaying bloom, decreases the potential for cold damage to blossoms and fruit. However, pruning very early, in late summer before the onset of rest, can stimulate late growth that will fail to harden and can winterkill. Early-fall pruning also delays hardening and interferes with flower bud formation by removing leaf surface and decreasing the supply of carbohydrates. In areas with extreme winter cold and little snow, winter damage could be severe.

In Florida, normal pruning and rejuvenation are done right after harvest, to allow another growth flush before frost (Lyrene, personal communication). Growers should experiment with pruning times carefully before settling on the proper time for their areas and cultivars.

SEVERITY

Blueberries should be pruned quite heavily. The tendency is to prune too lightly, because it removes flower buds and potential fruit.

The severity of pruning can influence ripening. Light pruning delays and prolongs the ripening season, while heavy pruning hastens and concentrates it (Brightwell 1941). Very heavy pruning can substantially decrease the yield of fruit per plant. Individual fruit size, however, will increase (Brightwell 1941; Brightwell and Johnston 1944). Heading-back cuts will reduce lankiness and willowy growth and stimulate the production of side shoots to the degree that they are made. Strong heading-back results in many laterals. These laterals will produce more flower buds and can increase production. Such buds bloom slightly later than those on main shoots. Weak plants should be pruned more severely than vigorous plants to stimulate growth (Mainland 1985).

DISTRIBUTION

Because the blueberry blooms only on one-year-old wood, each year's crop is borne farther from the root zone and the bush center. This means that nutrients must travel a longer distance to fruiting wood as well as supply much non-fruiting "extra" wood. Productivity substantially declines on canes older than five years, possibly due to a longer nutrient transport path. Older wood also conducts water and nutrients at an extremely slow rate, due to xylem clogging and crushing (Gough, unpublished data). Productivity declines on bushes with more than about eight to ten canes, probably due to competition. Therefore, all canes older than five years, all weak canes, and all excess canes should be removed annually. The size of the canes removed influences yield less than the percentage of the reduction in basal area of the bush (Siefker and Hancock 1987).

PRUNING YOUNG PLANTS

At planting, prune each cane to two to three vegetative buds. This will stimulate new growth strongly, as well as bring the top into better balance with the root system. Otherwise, top growth made during a moist spring can die back in a dry summer, without supplemental irrigation. A less severe practice for older, container-grown plants with larger root systems is to remove all weak, spindly growth and to head back shoots to remove flower buds.

After one growing season, the more vigorous bushes can be allowed to bear a small crop, less than 1 pt (0.5 l) per bush (20-30 flower buds). Remove any weak or unthrifty wood. If enough buds have not been removed, thin out or head back the remaining bearing shoots. On smaller, less vigorous plants, remove all weak growth (Figure 11-1a,b).

After two seasons, allow bushes taller than 2-3 ft (0.6-1 m) to bear a small crop of 1-2 pts (0.5-1 l) (Figure 11-2a,b). However, remove flower buds from tall canes that are not rigid enough to remain upright under a fruit load. The emphasis, however, should still be on establishing a healthy, vigorous bush and not on fruit production. Heavy production at this time can stunt the plants. Re-

Figure 11-1a. Vigorous 'Earliblue' bush after one growing season; before pruning.

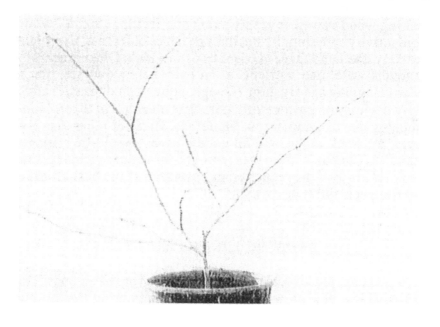

Figure 11-1b. Vigorous 'Earliblue' bush after one growing season; after pruning.

move all unthrifty growth. Follow a similar pruning procedure every year, allowing the plant to produce successively larger crops until the bush is mature. If well-grown bushes were started as healthy two-year-old plants, they areconsidered mature after six to eight growing seasons in the field.

PRUNING MATURE PLANTS

There are three points to consider when pruning a mature bush: prune lightly enough to ensure a heavy crop for the current year;

Figure 11-2a. Vigorous 'Coville' bush after two growing seasons; before pruning.

Figure 11-2b. Vigorous 'Coville' bush after two growing seasons; after pruning.

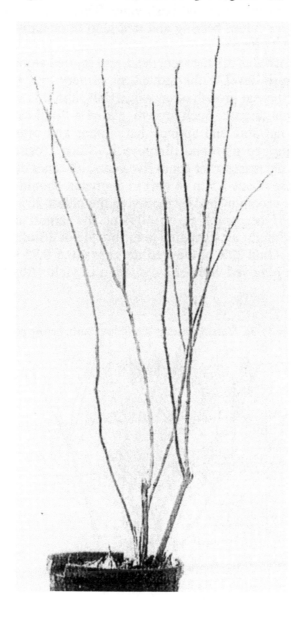

prune severely enough to secure large berries; and prune adequately enough to balance crop and bush vigor. This will assure sufficient new wood for future bearing and will help to contain overall bush size (Figure 11-3a,b).

After removal of all diseased, dead, and injured wood, cut old and unthrifty canes level to the ground, or to very low, vigorous side shoots. Soft, basal growth often winterkills, and it is very suscepti-ble to fungus disease. Such growth, called a "bull cane" or "bull shoot," is not stiff and springy but limber and often irregularly flat-sided instead of round. Remove it. Since a cane's profitable production decreases after about five years, all canes older than this, usually those about 1.5 in (4 cm) in diameter, should be removed. This can be accomplished by removing the oldest 20% of the canes each year. If necessary, up to 40% of the largest canes can be removed without substantially reducing yields (Siefker and Han-cock 1987). Only 20% of the medium canes (0.5-0.75 in or 1.25-1.9 cm) can be removed without a reduction in yield. Absence of new

Figure 11-3a. Mature, bearing 'Earliblue' bush; before pruning.

Figure 11-3b. Mature, bearing 'Earliblue' bush; after pruning. Note removal of some large canes and thinning of top.

canes frequently indicates that the bush is overcrowded with old canes and/or underfertilized. Complete removal of large and some medium canes will stimulate new growth. Prune these bushes severely enough to promote the growth of three to five new canes from the base of each plant each year. However, take care not to stimulate too many new canes through severe pruning, since this would necessitate additional work in removing them. It would also result in excessive vegetation shading the leaves. Shaded leaves are inefficient and use more food than they make. The most productive bushes are those with eight to ten canes, with 75% of those of medium thickness.

After the first two steps are completed, thin the bush by removing twiggy or bushy growth clusters and weak lateral shoots (Figure 11-4 and 11-5). This is particularly important with "bushy" cultivars, such as 'Northland' and 'Bluetta.' Erect-growing cultivars, such as 'Blueray,' 'Collins,' 'Bluecrop,' 'Herbert,' 'Earliblue,' 'Elliot,' and 'Jersey,' should be thinned out more in the center, while spreading cultivars, such as 'Coville,' 'Patriot,' 'Bluetta' and 'Weymouth,' usually require more pruning of the lower, drooping branches.

Since some plants tend to overbear, particularly in southern areas, thinning of flower buds can beneficial. This is especially true with cultivars such as 'Murphy' and 'Harrison' that can have up to a dozen flower buds on a shoot. It is accomplished by tipping back some of the bearing shoots to four to six flower buds. This thins the crop and improves berry size. Failure to thin the flower buds can yield large numbers of small fruit that fail to ripen or ripen very late. The number of flower buds to be left depends upon the vigor and bearing potential of the bush. Growers often leave about twice the number of buds needed for a good crop, to make up for those lost to frost, insects, and deer. After danger from these things has passed but before buds swell, excess buds can be removed. In pruning, always remove less vigorous, thin growth and leave thicker, more vigorous wood. Buds borne on thick wood open later in the spring and are thus less susceptible to late-spring frost damage. This type of wood also produces larger berries (Shutak et al. 1980). In North Carolina, flower buds formed in spring on first-flush growth will bloom and ripen fruit earlier than those formed in late summer, but they are also more likely to be damaged by bud mites and late frosts.

Figure 11-4. The bushy growth on the left of the figure should be removed during pruning.

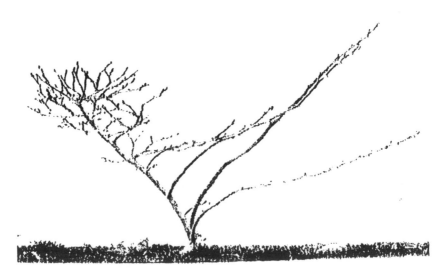

Figure 11-5. Short, lateral side shoots of this bush will produce inferior fruit and should be removed during pruning. This is done easily by sliding a gloved hand along the cane.

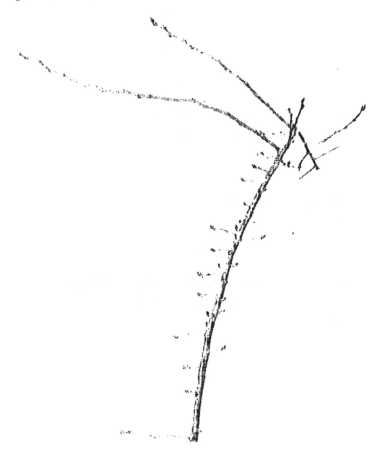

Unpruned bushes degenerate rapidly into a thick, twiggy mass of unfruitful wood.

REJUVENATION

In many instances, bushes are not pruned or are pruned insufficiently, resulting in overgrown plants with decreased vigor and production. Such plants may also be more prone to certain diseases.

Unless a change of cultivar is desired, or the bushes are diseased and must be removed, they can be rejuvenated and brought back into production by cutting back to the ground (Howell et al. 1975). The first summer after pruning there will be no crop, but bushes should bear a substantial crop the following summer and be in full production in the third year. One-half to one-third of the planting, depending upon the size, can be rejuvenated in consecutive years. This method is less expensive and causes a smaller crop loss than removing the plants and replanting. However, it results in all canes of the same age, which is not desirable, and in multiple, vigorous shoots that are highly susceptible to winter damage. An alternate method is to cut back half of each bush to stubs and allow the other half to bear fruit. The following year, the previously unpruned half is cut back. This procedure will completely rejuvenate a bush in two years, while providing a partial crop each year.

COST

Mainland (1989a) reported that, by using highly trained labor to make the first several large cuts, plants could be properly shaped and the time required for detailed thinning cuts could be substantially reduced (Table 11-1). Funt et al. (1991) reported on the time required to prune plants of several cultivars (Table 11-2). This is only a guide, since the true amount of pruning time will depend to a great extent upon the condition of the bush and the experience of the pruner.

Table 11-1. Cost of hand-pruning highbush blueberries.

Item	Cost per acre
Labor–40 hr per acre @ $5.25 per hr	$210.00
Chop prunings	
Labor–1 hr per acre @ 5.25 per hr	$ 5.25
Equipment–Tractor @ $5.66 per hr	
Mower @ $2.96 per hr	$ 8.62
	$223.87

Source: Mainland 1989b

Table 11-2. Time required to prune several cultivars of highbush blueberry.

Cultivar	Time to prune (min.)[z]
Blueray	1.52
Bluecrop	1.20
Berkeley	1.67
Jersey	1.98
Herbert	1.89
Coville	1.12
Lateblue	1.73

[z]Average of 10 mature bushes in minutes per plant.
Source: Funt et al. 1991.

MECHANICAL PRUNING

Interest in mechanical pruning has increased recently due to the difficulty encountered in finding enough skilled hands who are willing to work. In North Carolina, plants are cut back to a height of 4 ft (1.2 m) immediately after harvest by using a sickle mower (Mainland 1989b). With sufficient rain (or irrigation) and good vigor, plants will develop enough flower buds for the next year. This summer mowing will reduce dormant hand-pruning by more than 50%. Dormant pruning is then reduced to removal of old, poorly positioned, and crowded canes. Most fruit are borne on new growth that develops above the sickle cut. Because annual summer topping will eventually reduce yield, the procedure is recommended to be performed only every other year.

In addition to reducing bush height for easier harvest, old top growth that may be infested with *Phomopsis* or mites is removed and new flower buds are formed well above the remaining old growth. However, the new growth that develops following topping is attractive to sharp-nosed leafhoppers, requiring close attention to a spray schedule.

Chapter 12

Propagation

Although most growers are not financially and physically able to undertake large-scale propagation operations, they may benefit from small-scale propagation of particularly valuable cultivars or those cultivars that are difficult to procure through nurseries specializing in blueberry propagation.

The highbush blueberry is most commonly propagated by hardwood cuttings, though softwood and semihardwood cuttings can also be used; still other methods include grafting, seedage, mounding, clump division, layering, and tissue culture.

MEDIA

The media for rooting hardwood and softwood cuttings should be well drained but moisture retentive. An equal mixture, by volume, of peat and vermiculite or coarse sand is excellent. Moore and Ink (1964) reported that increasing the ratio of acid peat to sand from 1:1 to 2:1 increased the rooting of cultivars that are notoriously difficult to root, such as 'Bluecrop.' Shelton and Moore (1981a) found that straight peat gave excellent rooting results but that the rooted cuttings were difficult to separate. The media should be placed in the tray or bed about a week ahead of time and thoroughly soaked. Hot water may be necessary to wet dry peat. Never add fertilizer to the media at this time.

BEDS

Beds for cuttings can be made in any convenient size, with sides 6-8 in (15-20 cm) high. Place galvanized screening on the bottom to exclude grubs, moles, and mice and to provide good drainage and aeration. Set the bed on coarse gravel, cedar logs, or cement blocks to keep the bottom off the ground. Electric bottom heat can be installed at this time and regulated to keep the media as close to 70°F (21°C) as possible. Protect cuttings in trays from direct sunlight and excessive drying by placing them in a lath house or under a shade of cheesecloth or tobacco cloth. An arch of concrete-reinforcing wires covered with opaque white polyethylene, shade cloth (45-50% shade), or burlap may be substituted. If polyethylene is used, remove it as soon as most of the cuttings have rooted, but leave the shade on until the end of summer. In larger frames, keep cuttings moist with an automatic mist system. This system has given good results without any additional shading. In smaller frames, hand-watering may be in order. Never hand-water during the hottest part of the day or in the evening.

HARDWOOD CUTTINGS

These are taken from dormant, healthy, well-matured shoots of the previous season's growth. Avoid poorly hardened or thin (less than pencil thick) shoots. Take cuttings only from mother blocks certified free from virus. Shoots from which cuttings are taken are usually 10-30 in (25-75 cm) in length and are called "whips." Use only the portion bearing leaf buds. The region of the shoot from which the cuttings are taken is important because of the content of various nutrients and hormones in different portions of the shoot. A low-nitrogen, high-carbohydrate balance in various shoot regions seems to favor rooting in many plants. This situation apparently exists near the base of many shoots. Other compounds conducive to rooting may also exist in optimal proportions near the base. The formation of flower buds near the shoot tip apparently alters some physiological condition in the shoot, possibly auxin distribution, that decreases its rooting capacity. Simple removal of the flower buds will not substantially increase the rooting capacity of the shoot.

Since roots initiate near the center portion of a cutting, shoots with a brown pith, which indicates injured tissue, will not root well.

Wood for cuttings can be taken after leaf drop in the fall or before new growth begins in the spring. If they cannot be planted right away, they can be stored at 36°F (2°C) in moistened, sealed poly bags or packed in damp sphagnum moss, sawdust, or similar material. This will satisfy the chilling requirement of the buds. Wood may also be taken in the spring before bud growth begins. However, Shelton and Moore (1981b) reported that cuttings collected in fall or early winter were better than those collected after midwinter, provided their chilling requirement had been met.

Cuttings are usually 4-5 in (10-13 cm) long and about pencil-thick. Thicker cuttings do not root well. Use the basal or medial portions of shoots (Moore and Ink 1964). Make the bottom cut just below a bud and the top cut about 0.25 in (0.6 cm) above a bud. All cuts should be clean, with no crushed tissues. Cuttings are usually placed vertically in the media 2 in (5 cm) apart, with 2 in (5 cm) between rows and with only the top one or two buds showing. Pack the media firmly around the cutting. The success of treatment with a root-promoting chemical is dependent upon the hormone and the cultivar (unfortunately, this type of treatment has not generally been successful). Nevertheless, Hormodin #3 has been useful in improving the rooting of 'Bluecrop' and Hormo-Root C in improving the rooting of 'Blueray' and 'Stanley.'

After setting the cuttings, thoroughly water to settle the media around them. Leaves on the cuttings will develop quite early and maintenance of high humidity through periodic misting is essential until roots form after the first flush of top growth is complete. This may take two to three months. Shelton and Moore (1981a) found that propagating in full sun or in an area with up to 50% shade gave best results. More than 50% shade was detrimental to the rooting process.

When the second flush of top growth begins, roots have formed and fertilization is recommended. A high-analysis liquid fertilizer is used at the rate of about 3 lb (1.3 kg) in 50 gal (190 l) of water. One gal (3.8 l) should be enough for about 25 sq ft (2.25 sq m) of bed surface. Rinse all foliage with water immediately after application. Weekly applications are made until late summer.

Rooted cuttings may be kept in the propagation bed all winter, transplanted to the nursery row, or potted in early fall, preferably by mid-September in colder regions. When plants remain in beds through the winter, they should be protected by a heavy mulch and soil mounded around the sides of the frame, to reduce bottom ventilation and freezing.

When cuttings are transplanted to the nursery row, set them about 1 ft (30 cm) apart in rows spaced about 18 in (50 cm) apart, and apply a good mulch. If soil is fertile, no fertilizer is needed the first year. After one year in the nursery, plants are considered two years old. Cuttings to be transferred to pots are usually placed in a standard 1-gal (3.8-l) nursery container, holding about 3 qt (2.83 l) of soil. They may also be lined out 6 in (15 cm) apart in propagation beds and grown on for one year.

SEMIHARDWOOD AND SOFTWOOD CUTTINGS

These are taken in the early summer while the plant is actively growing. Young plants give cuttings with the highest percentage of rooting. The best time to gather them is just after the first flush has hardened sufficiently to handle and just before the second flush has started (Johnston 1935; Douglas 1967; Coorts and Hull 1972). Cultivars that ripen their fruit earlier–'Earliblue,' 'Collins,' 'Bluecrop,' and 'Blueray'–are somewhat easier to root than those that ripen fruit later, such as 'Berkeley' and 'Coville.' Also, 'Concord,' 'Herbert,' 'Stanley,' 'Ivanhoe,' and 'Bluecrop' are often easier to root by this method than by hardwood cuttings (Doughty et al. 1981). Cuttings should be 4 in (10 cm) long, with all the leaves but the upper two removed and collected not later than mid-morning on sunny days to avoid excessive transpiration. Heel type (Schwartze and Myhre 1954) and leaf bud cuttings (Parliman et al. 1974) can also be used. The latter is particularly useful when propagation material is in short supply. When mist is used, remove only the leaves on the lower half of the cutting. The physiological bottom of the cuttings can then be moistened and dipped in a rooting compound, such as a talc formulation of 4.5% indolebutyric acid (IBA), to promote rooting. Coorts and Hull (1972) reported that 0.8% IBA plus 15% tetraethylthiuram disulfide (Hormo-Root C) gave good results. Cuttings should be inserted into a media of 1:1 peat-to-perlite down to the lowest remaining leaf. Pinebark medium is not as beneficial, since daily watering readily leaches the already low nutrient content (Lyrene, personal communication). The care of these cuttings is similar to that of hardwood cuttings, except that high humidity and ventilation are even more critical. Mist beds delivering about 15 seconds of mist every hour are strongly recommended. Do not allow leaves to dry. Periodic fungicide applications–misting only during daylight hours–and good ventilation will help control tip blight and rotting of the succulent stem and leaf tissues caused by *Botrytis*. Ventilate to prevent moisture from condensing on the foliage, but not enough to cause wilting. Ferbam or Benlate will control severe *Botrytis* problems. Treat the entire bed, including the frame, weekly for a month.

Softwood cuttings will root well in up to 60% shade, but where the propagating room has no windows, cuttings will root under a combination of fluorescent and incandescent lights delivering 100-200 footcandles of light (Waxman 1965). Intensities approaching 300 footcandles result in foliage discoloration and poor rooting. After rooting, the cuttings can be fertilized with weekly applications of a water-soluble fertilizer.

Other methods of propagating highbush blueberry include grafting, seedage, mounding, and tissue culture. The first approach is difficult and can be used only in limited situations. Ballington et al. (1989a) reported on the potential commercial value of grafting highbush blueberry onto rootstocks of *Vaccinium ashei* and *V. arboreum*. They found spring stub grafting and saddle grafting, followed by hot callusing in mid-winter, both to be successful. The second is useful for plant breeders only, since the blueberry does not come true to seed, and the last is too elaborate for most growers. Read et al. (1989) found that 'Northblue' plants produced through tissue culture generally had a bushier habit, more vigorous shoot growth, and greater flower bud formation, resulting in higher yields than plants produced by conventional methods. Grout et al. (1986), also working with 'Northblue,' found that plants produced by this method grow up to three times faster than plants produced from leaf-bud cuttings. Rowland and Ogden (1992) reported that supplementing the culture medium with zeatin riboside resulted in greater shoot regeneration from leaf sections than from sections placed on 2iP medium.

Chapter 13

Fruit Production

The blueberry flower resembles an urn tipped upside down. The corolla is usually white, but may be tinged with pink, particularly during cool weather. Flowers of 'Blueray' are usually pinkish white and those of 'Northland' are greenish white. Inside the corolla and near its base are the nectar-secreting glands, or nectaries.

Because of genetic, climatic, and cultural factors, a blueberry plantation of many cultivars can be in bloom for as long as a month. With the exception of 'Duke,' 'Spartan,' and 'Bluejay,' cultivars that ripen their fruit early also bloom early (Hancock et al. 1991). Plants of earlier blooming cultivars have a greater number of days with open flowers and a longer period of full bloom than those of later-blooming cultivars, since those cultivars that bloom earlier are doing so under generally cooler temperatures and thus have a longer bloom period (Gough et al. 1983). In addition, the entire bloom period can be shifted forward in cool springs or backward in warm springs. Gough (unpublished data) found that 'Earliblue' reached full bloom when approximately 1000 growing degree days (base 43°F or 4.4°C) had accumulated since the beginning of the year. Other factors—such as plant vigor, pruning time, mulching, and flower bud position on the plant—also have an influence. The flower bud at the tip of a shoot generally opens first, followed by the one just beneath it, and so on down the shoot. The opposite sequence occurs within each bud, with the flower near the base of the cluster opening first (Hindle et al. 1957). Blossoms on thin wood open before those on thick wood (Hindle et al. 1957). Weak plants, therefore, can be more susceptible to spring frost damage. Although there appears to be no correlation between flower bud position on the bush and time of bloom, buds near the bottom of plants mulched with sawdust will be among the first to open, due to the generally warmer air temperatures in this area caused by the mulch (Gough, unpublished data). Plants pruned in the fall generally bloom a few days later than those pruned in the spring (Gough 1983c). Poor weed-control practices, insufficient watering, sandy soil, improper fertilization, and improper pruning can also alter bloom times slightly, since all could result in weak shoot growth (and weak shoots bloom early). With so many factors influencing the precise time of bloom, it is difficult to predict exactly when plants of a certain cultivar will flower. However, by averaging the dates of full bloom of various cultivars in your locale over a number of years, a reasonably accurate prediction can be made.

Internal flower bud activity increases with the first warm days of spring and is evidenced by external swelling. Within one to one-and-a-half months before bloom, the areas that will accommodate seed development in the ovaries have begun their final phase of development. Pollen maturation in the anthers has begun and pollen grains appear to be fully formed and ready for shedding a few weeks before bloom. A couple of days before bloom, the style begins to elongate, pushing the corolla into a long cylinder. Finally, the corolla opens, exposing the pollen-bearing anthers and the stigma to foraging bees. Highest fruit set is associated with the shortest styles (Table 13-1) (Eck and Mainland 1971).

Table 13-1. Fruit set relationship among some cultivars of blueberry.

Relative set	% Set	Cultivars
Excellent	86-100	June, Rubel, Bluecrop, Rancocas, Murphy, Angola
Good	71-85	Cabot, Wolcott, Croatan, Darrow, Ivanhoe, Burlington, Pemberton, Dixi, Concord, Weymouth, Blueray
Fair	56-70	Atlantic, Berkeley, Collins, Jersey, Herbert
Poor	<56	Stanley, Coville, Earliblue

Source: Eck and Mainland 1971.

Blueberries are usually pollinated by insects, primarily wild bumblebees and the domesticated honeybee. Although bumblebees are the greatest natural pollinators of blueberries (Merrill 1936), sufficient wild populations are often lacking (Filmer and Marucci 1964). As a result, placement of honeybees in the field will nearly always increase yield. Also, when the bee population is high, the more attractive flowers are pollinated quickly, forcing the bees to work sooner on those less attractive. This makes the entire pollination process more efficient. Use at least two to three strong colonies of bees per acre (0.4 ha) of plantation. A weak colony with three frames of brood and 15,000 bees has only 40% of the pollination potential of a strong colony with six frames of brood and 30,000 bees (Lyrene, personal communication). Marucci (1966) recommended one colony per 2 acres (0.8 ha) of attractive cultivars such as 'June,' 'Rancocas,' and 'Rubel' and one colony per acre (0.4 ha) of 'Weymouth' and most other cultivars. Two colonies per acre (0.4 ha) should be used on less attractive cultivars, such as 'Coville,' 'Stanley,' 'Berkeley,' 'Jersey,' 'Elliot,' '1613-A,' 'Concord,' and 'Earliblue' (Pavlis 1991a). These should be brought into the field as soon as the first plants of the earliest cultivars reach full bloom (Howell et al. 1970, 1972); they should be removed before insecticide spraying begins. Place hives near the center of the plantation, facing east, in full sun, and with a windbreak provided where necessary. Effective flight distances usually do not exceed 300 yd (300 m). Recently, Corliss (1992) reported that *Osmia ribifloris*, a wild bee found in the coastal mountains of Southern California was more efficient in pollinating blueberries than honeybees. The bees visit blossoms about every three seconds–three times faster than honeybees, and fly in poorer weather conditions. MacFarlane (1992) reported that *Bombus terrestris*, the short-tongued bumblebee, and the honeybee were the most important pollinators in New Zealand.

Be sure to eliminate nearby plants that bloom with the blueberry, such as dandelions. Close mowing or the use of herbicides will prevent them from competing with blueberry flowers for bee visitations.

The problem of deciding whether or not to rent hives can be difficult. In general, expect adequate pollination if you observe 15-20 bee entries into blueberry flowers within a ten-minute period.

Pollinated flowers turn upright, remain white, and drop within four to five days. Flowers that have not been pollinated within the critical few days after they open remain on the bush from 9-12 days and often turn a brilliant wine color. 'Northland' in New England displays this more than other cultivars. Ovaries that have not been fertilized will turn reddish or yellowish-red and drop without substantial swelling. These are termed "redcaps." Some drop is normal but large numbers of fallen, discolored ovaries indicates poor pollination. Fruit set with good pollination will swell rapidly and will contain, on average, 30 seeds each.

Even if the grower has provided for adequate cross-pollination, the crop in some abnormal springs may be small. Bees may not fly in hot, rainy, drizzly, or windy (above 15 mph) conditions or in temperatures below about 55°F (13°C). The optimum temperature range for bee foraging is 65°-70°F (18°-21°C). If temperatures drop to near 40°F (4.5°C) after pollination, the pollen may not germinate or the tube may grow so slowly that fertilization does not occur. Under very humid conditions, pollen grains can clump together and not be carried by bees. Very dry conditions also do not favor pollination.

If temperatures fall below freezing, flowers can be damaged, making fertilization impossible. Such damage is not always obvious at first glance. For example, the stigma and style can be damaged and browned, making fertilization impossible, but the corollas may appear sound. Depending upon cultivar and preceding temperatures, flowers in full bloom can be damaged at temperatures of about 28°F (−2°C) for four hours duration or 26°F (−3°C) for two hours. In general, if the temperature remains above 28°F (−2°C) in an official slatted thermometer housing 4 ft (1.3m) off the ground, flowers will not be harmed. After petal fall, critical temperatures for injury rise to 29°−30°F (−1.6° to −1.1°C), since berries lose heat faster than flowers (Lyrene, personal communication). If the temperature drops rapidly from relatively mild temperatures, injury can occur at even higher temperatures. If cool weather prevails during bloom, flowers may not be damaged until the lower temperature is reached. Since earlier-ripening cultivars usually bloom earlier than later-ripening ones, and therefore may be more apt to experience colder temperatures, plant them near the top of a slope to provide the best air drainage and fewer temperature fluctuations.

To assess the extent of frost injury, wait two days after the suspected damage occurred and slit open the ovaries and seed chambers. These will be bright green to white in undamaged fruit, brown to black in damaged fruit.

Bees are attracted by both nectar and pollen, and some cultivars are more attractive than others (Filmer and Marucci 1964). The precise reason for this is unknown. Flowers of 'Earliblue,' 'Coville,' 'Stanley,' 'Dixi,' '1613-A,' 'Berkeley,' 'Concord,' and 'Jersey' are relatively unattractive to bees. Those of 'Rubel,' 'June,' and 'Rancocas' are very attractive. Dense plantings of unattractive cultivars should be saturated with bee activity and provided with good cross-pollination conditions. Flowers on winter-damaged wood appear small and yellowish white and are unattractive to bees.

In recent years, several compounds have been marketed that are purported to attract honeybees to flowers. These materials are not particularly effective in attracting bees to blueberry flowers.

Because of the structure of the blueberry flower, most bees will have to brush the pollen-bearing anthers to reach the nectar secreted by the nectaries near the floral base. When this happens, pollen sticks to the bee and is carried along from flower to flower. Fortunately, the bee must also brush the stigma, to which some of the pollen sticks. This process is sometimes circumvented. Carpenter bees and a species of bumblebee, *Bombus occidentalis occidentalis* Greene, bores a small hole through the corolla near the base of the flower, most often in 'Bluecrop' and 'Jersey,' and gathers the nectar without touching either the anthers or the stigma (Figure 13-1). Honeybees also may then use the holes (Eaton and Stewart 1969). Further, if corollas are opened very wide, as in 'Earliblue,' the bee can insert its tongue down to the nectary without touching the stigma.

Blueberries can set fruit by self-pollination or cross-pollination, depending upon the cultivar. Although Lang et al. (1990) reported that absolute self-incompatibility was not apparent in *Vaccinium corymbosum*, self-pollination is generally not desirable. 'Jersey,' 'Weymouth,' 'Northblue' and other cultivars sometimes form parthenocarpic fruit. These have no fertilized eggs or seeds, do not

Figure 13-1. Photomicrograph of a longitudinal section of a blueberry ovary with flower attached during bloom. The sepal protrudes to the left of the figure. What appears to be a forked structure near the center is the corolla (to the left) and filament (to the right). The darkly stained ridge to the right of the figure and at the top of the ovary is the nectary. Bumblebees often bore through the base of the corolla and gather nectar here–instead of entering the flower near the top–brushing the anthers and accomplishing pollination. (Source: Lohbusch and Gough, unpublished data).

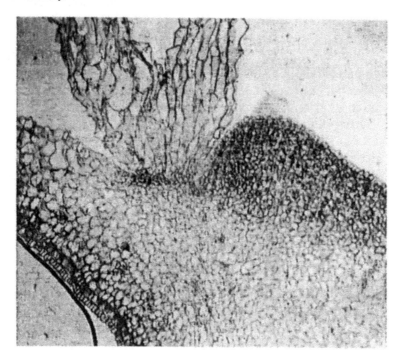

develop properly, and remain unmarketable. In a later study, Lang and Parrie (1992) reported that 'Avonblue' had significantly lower pollen germination than 'Georgiagem,' 'Flordablue,' 'Gulfcoast,' 'O'Neal,' and 'Sharpblue,' all of which had greater than 90%.

Most highbush blueberry cultivars are self-fruitful–setting fruit by their own pollen (Merrill 1936)–but they will benefit from cross-pollination (Vander Kloet 1991; El-Agamy et al. 1981; Meader and Darrow 1947; Bailey 1938; Beckwith 1930). Berries will ripen earlier, be larger and more plentiful, and contain more seeds if ovaries have been cross-pollinated (Vander Kloet 1991; Lang and Danka 1991). Lang and Danka (1991) found that cross-pollination of 'Sharpblue' did not increase the number of fruit set but did improve fruit quality. Some cultivars, such as 'Stanley,' 'Coville,' 'Elliott,' 'Northcountry,' 'Northsky,' 'St. Cloud,' and 'Earliblue,' are partially self-incompatible. They will not set good crops by their own pollen. In addition, much 'Coville' pollen is defective and is shed in only small amounts. Three or four interplanted cultivars will improve cross-pollination. In Florida, interplanting 'Sharpblue' with 'Misty,' 'Flordablue' or 'Gulf Coast' is particularly effective. In larger plantings, no cultivar should be separated by more than two rows from a cultivar with a similar bloom period. In general, the time of bloom roughly corresponds with the time of ripening: the early-ripening cultivars, for instance, also bloom early.

The stigma and egg remain receptive to the pollen and sperm for only five to eight days after bloom (Moore 1964). However, if pollination does not occur within about three days after the flower opens, fruit set is unlikely (Merrill 1936). Because of this, saturating the planting with bees early in the bloom period is necessary for adequate fruit set. Once pollen has been deposited upon the stigma, it usually will germinate, allowing the sperm to fertilize the egg.

When temperatures are warm, pollen often germinates immediately after landing on the stigma, and the pollen tube begins to grow down the style. Upon reaching the egg sac (which may take one to two days) the tube ruptures, two sperm are released, they fuse with female nuclei, and fertilization is completed. When this occurs, the outer coating of the ovule begins to harden into the seed coat. Because a blueberry fruit can have up to several dozen seeds, with each seed requiring a pollen grain for its development, the importance of good pollination is obvious. Further, there is a correlation between the number of developed seeds and fruit size. With better cross-pollination, more seeds develop, resulting in larger fruit. Additional factors also influence fruit size.

Following fertilization, an increase in the amount of certain plant hormones (including gibberellins) in the ovary itself causes rapid enlargement of the ovary and development of the ovules into seeds. Once this happens, the corolla fades quickly and drops, leaving the swollen ovary "set" and beginning to develop into a fruit.

The problem of deciding whether or not pollination activity has been adequate is often difficult for the grower. If the number of unpollinated flowers exceeds 20% of the total, hormone sprays are indicated. Some New Jersey growers routinely spray cultivars that are unattractive to bees. Synthetic gibberellins can be applied to the flowers. They stimulate the ovary tissue directly and make pollination unnecessary. Because the ovules were not fertilized and developed into seeds, the resulting fruit will be parthenocarpic, or seedless. Gibberellic acid sprays may also cause more fruit to set than might be the case under normal conditions. This will increase competition among developing fruit and, along with the fact that they are seedless, will cause them to be somewhat smaller in size and later-ripening (Mainland and Eck 1969). These sprays also stimulate the sizing of "shot" berries which normally ripen, but are usually very small and seedless. Sizing makes them more marketable. Shot berries are often found on 'Jersey.' Because of these problems, do not use such sprays when normal pollination is adequate. A single application of 200-ppm gibberellic acid can be applied from 0-14 days after 75% of the flowers are in full bloom. Wet foliage to the point of runoff, do not exceed 300 gal of spray per acre, and do not apply the product within 40 days of harvest. Follow the manufacturer's directions carefully. The compound dissipates quickly in fruit. If used too early in the bloom period, it can result in an excessive number of late-ripening, undersized berries. Because these compounds will often cause more fruit to set than would occur under normal conditions, they should not be used on very young bushes or on bushes low in vigor. Excessive fruit set under these conditions can result in substantial stress to the bush, which could influence future growth and productivity.

It sometimes is necessary to speed and concentrate ripening and to increase the ease with which berries separate from the bush. Ethephon (2-chloroethylphosphonic acid) is registered for use on blueberries and will accomplish this. Response of the fruit to this spray will vary according to climate, cultivar, and timing. Follow the manufacturer's directions carefully.

Chapter 14

Harvesting

Blueberry fruit require two to three months to ripen, depending upon weather, the type of wood left after pruning, and the cultivar. Many complex biochemical changes occur within the fruit during this time, including those which result in softening, heightened pigmentation, sweetening, and enlargement. If berries are harvested as soon as they show a slight pink coloration, the ripening process will still continue, but fruit quality will be lower than if the berries had ripened on the bush.

The sugar content of a blueberry on the bush ranges from 7% in a green berry to about 15% in a ripe berry (Shutak et al. 1980). Most of this sugar is manufactured in the leaves and transported into the fruit. Consequently, berries ripened off the bush often reach a final sugar content of only about 10%. Fruit left on the bush will continue to increase in sugar content even after they turn blue. Berries harvested as soon as they turn blue contain about 12% sugar. If allowed to remain on the bush for an additional three to five days, they can attain up to 15%.

The blueberry fruit is about 83% water. Because it can only receive this through the plant, a berry cannot increase in size substantially after harvest. Depending upon the cultivar, a delay in harvest of six days after the first appearance of color can mean up to an additional 382 qt per 1000 qt of berries (420 l per 1100 l) (Shutak et al. 1957). Drought during ripening reduces the amount of water available for bush and berry growth and results in (1) smaller, fewer leaves (which are less able to supply adequate sugar for full berry flavor) and (2) less water being available for increases in fruit size.

Highbush blueberry harvest in the United States begins in April with 'Flordablue' in Florida and continues until about mid-September with 'Elliott' in northern areas. Carlson and Hancock (1991) developed a model for determining the heat unit requirements necessary for harvest. They found it to be from 22-69% more accurate than calendar date. However, variation was highly dependent upon cultivar. One hand-picker can harvest 60-80 pt (33-44 l) per eight-hour day. Each acre (0.40 ha) should be allotted two pickers at the beginning and end of the season and up to eight pickers per acre during peak harvest (Baker and Butterfield 1951).

Because the fruit does not ripen simultaneously in the cluster, the entire plantation should be picked about once per week for up to a half-dozen weeks, depending upon the cultivar. The largest fruit will ripen during the first two pickings. Ripe berries should be rolled from the cluster into the palm of the hand with the thumb. This reduces the harvest of immature berries and tearing and bruising of the fruit. A shoulder or belt fastener can be provided for the container so that both hands are free for picking.

Berries should be placed directly into the boxes in which they will be sold. It is important that there be a minimum of handling so as not to destroy the berries' attractive, whitish surface bloom or to increase bruising and spoilage.

Berries for fresh market should be plump, firm, uniformly blue, and free of injury and trash. They usually are packed in plastic, veneer, paperboard, or pulp containers with a 1 pt (0.5 l) capacity. These are covered with cellophane held in place with a rubber band. The film reduces water loss, keeps out dust, and improves the appearance of the fruit. These containers are then shipped in larger wooden crates holding 12 1-pt (0.5-l) boxes in a single layer. The larger containers require about 403 cu in (6600 cu cm) and weigh 7-9 lb (3-4 kg) (Ryall and Pentzer 1974). Most blueberries produced in the southern United States and in New Jersey are harvested by hand for fresh market (Hancock and Draper 1989).

A quarter-century ago, mechanical harvesters were praised as the thing of the future. Although the technology of such harvesters has improved greatly, problems still plague their use. Fruit loss–including dropped, bruised, green, and stemmy berries–has been estimated at 25-50% (Hansel et al. 1971; Liner 1971; Mainland 1971). Howell et al. (1975) reported that stem damage from mechanical harvesters was associated with increased winter injury. Pruning bushes to 9-in

(23-cm) bases and modifying the harvesters somewhat can reduce some damage (McCarthy and Rohrbach 1985). More recently, Rohrbach and Mainland (1989) reported that crown-restricting collars made of agricultural drainage tubing and placed over young plants constricted bush structure at the base and reduced mechanical harvesting ground losses by 40%. Mainland et al. (1975) estimated that mechanical harvesting caused 50 times more injury to the bush than hand harvesting and also resulted in 10-30% softer fruit and 11-41% more decay. Despite these problems, and because of the increased cost of laborers, nearly the entire Michigan crop is harvested by over-the-row machines for the process trade (Burton et al. 1979; Hancock and Draper 1989).

The suitability of different cultivars to mechanical harvesting also varies (Galletta and Mainland 1971). 'Bluecrop,' 'Collins,' 'Coville,' and 'Morrow' showed similar suitability for fresh market (whether hand- or machine-harvested), while 'Earliblue,' 'Murphy,' and 'Wolcott' were not suitable when harvested by machine. Further, mechanical harvesting increased the decay in 'Bluecrop' by 23-33%, presumably due to bruising and mechanical damage.

Mechanically harvested fruit cost the grower about $0.18 per lb (0.45 kg) and, because of the lower quality as a result of the harvesting procedure, the fruit are primarily sold for processing at around $0.50 per lb. On the other hand, it costs the grower about $0.41 per lb, including packing and labor, to hand-harvest fruit for the fresh market, but he will realize a higher selling price of about $1.00 per lb to consumers (Safley 1985).

After harvest, berries are sorted by hand or by machines whose operations are based upon sorting by firmness (vibration, bouncing, etc.), mass density (sink/float, air blasts), or optical density. One of the most dependable, least expensive, and perhaps simplest method is the air blast system. The less dense leaves and trash are blown away from berries falling through a rapidly moving stream of air.

Respiring berries release heat. Since respiration is a chemical reaction, the warmer the fruit, the faster they respire. The process uses sugars and other compounds in the berries, eventually resulting in loss of quality. Cooling as soon as possible to remove field heat and slow respiration will significantly prolong the quality and shelf life of the fruit by conserving sugars and retarding production of ethylene and growth of decay organisms. In fact, decay is the most serious consequence of high holding temperatures (Ryall and Pentzer 1974). At 32°F (0°C), a ton of blueberries releases 500-2300 Btu (126-580 kcal) per day. This jumps to 200-2700 Btu (300-675 kcal) at 40°F (4.4°C); 7500-13,600 Btu (1875-3427 kcal) at 60°F (15.6°C); and 17,200-27,300 Btu (4334-6880 kcal) at 80°F (27°C) (Lutz and Hardenburg 1968). Fruit in mesh and plastic containers cool faster than those in pulp and cellophane (Rohrbach et al. 1984). Berries held in the field in bright sun rapidly lose quality. Keeping quality is also dependent upon cultivar (Shutak et al. 1980). Cultivars displaying higher respiration rates after harvest are least likely to keep well (Figure 14-1).

Sound, ripe fruit free from moisture can be held for up to two weeks under the optimal storage conditions of 31°-32°F (−0.5° to

Figure 14-1. Correlation between peak respiration rate and keeping quality of seven cultivars of highbush blueberry. (Source: Shutak et al. 1980).

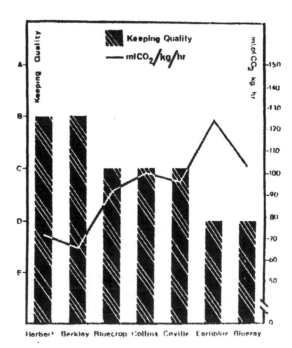

0°C) and > 90% relative humidity without significant loss of quality. Storage for up to four weeks under these conditions will result in some quality loss and off-flavors. Storage at normal home refrigerator temperatures of around 40°F (4.4°C) results in substantially greater wastage than at 32°F (0°C). Recently, work by Song et al. (1992) and Beaudry et al. (1992) has shown that blueberries can be kept in modified atmosphere storage, though this method is not yet commercially feasible. Prang and Lidster (1992) reported that controlled atmosphere storage substantially reduced live maggot infestations in lowbush blueberry without adversely affecting fruit quality.

As mentioned above, slow cooling and mechanical damage can increase the incidence of post-harvest decay. As many as 16 different fungi have been found on blueberries after harvest (Cappellini et al. 1972). The most important of these are described below.

Gray Mold

Causal Organism: Botrytis cinerea

Description: Grayish mycelium develops on the fruit surface and berries undergo a soft, watery breakdown.

Development: Infection occurs before harvest, especially during wet periods. The organism grows slowly in storage and spreads by contact. Decayed fruit often clump together. It commonly infects fruit harvested later in the season.

Prevention: Follow a good fungicide program and grow resistant cultivars.

Anthracnose

Causal Organism: Gloeosporium sp.

Description: The surface of the fruit is covered with small, black fruiting bodies. Under moist conditions, fruit may show masses of salmon-colored spores.

Development: Infection develops slowly in the field and only a few cultivars are susceptible.

Prevention: Follow a good fungicide program.

Mummyberry

Causal Organism: Monolinia vaccinii-corymbosi

Description: Immature green fruit shrivel before harvest and a cream-colored mycelium develops on their surface.

Development: Pre-harvest. The disease will not spread from berry to berry.

Prevention: See section on disease control.

Rhizopus Soft Rot

Causal Organism: Rhizopus nigricans

Description: Juice oozes from disintegrating fruit. At temperatures above 70°F (21°C), a heavy growth of white mold develops on the fruit surface.

Development: Infection is through mechanical injury and is spread by fruit contact. Presence of free water favors decay.

Prevention: This rot causes more market losses in berries than any other disease. Since the organism will not grow at temperatures below 50°F (10°C), prompt cooling and careful handling will help control the disease.

Vigorous plants start producing the second year after planting and continue in well-managed plantations for more than 50 years. Yields vary substantially, depending upon overall climate, weather patterns, soil conditions, cultivar, and management practice, but will commonly average about 3.8-4.6 lb (1.5-1.8 kg) on mature plants. On well-managed sites, yield can increase up to 20-24 lb (8-9.6 kg) per bush. These translate to 1.9-2.3 tons per acre (4.5-5.5 metric tons per ha) and 10.3-12.3 tons per acre (25-30 metric tons per ha), respectively (Hancock and Draper 1989). Yields in excess of these, sometimes as high as 64 lb (29 kg) per bush, have been reported (Funt et al. 1991). Funt et al. (1991) also reported on the yield of highbush blueberry under experimental conditions in Ohio (Table 14-1). They separated yields into early and late, and determined that high early yields of more than 2500 lb per acre per year (2800 kg per ha) were necessary for an acceptable rate of return on investments. Later yields should average more than double this figure.

Table 14-1. Early and late yields of highbush cultivars in Ohio.

| Cultivar | Total yield per plant[z] | |
	Early[y]	Late[x]
Bluetta	7.4	15.1
Harrison	2.4	8.3
Collins	6.7	13.2
Northland	5.6	15.6
Spartan	4.2	11.0
Bluejay	7.5	17.6
Patriot	3.6	3.6
Bluehaven	0.9	4.3
Blueray	7.8	19.4
Bluecrop	4.6	16.2
Berkeley	6.3	14.2
Jersey	4.6	13.2
Herbert	10.7	16.8
Darrow	7.6	17.5
Coville	3.4	15.8
Lateblue	7.4	28.8
Elliott	6.4	21.5

[z]Cumulative yield in kg.
[y]Yield in first four harvest years.
[x]Yield in last four harvest years.
Source: Funt et al. 1991.

Chapter 15

Pest Control

BIRDS

The pests most injurious to blueberry production are birds. They are especially destructive in small plantings located in, or near, highly populated areas; with increased urbanization, the problem is becoming more severe. Birds not only steal or damage ripening fruit but can destroy flower buds as well. If this trend continues, blueberry plantations may have to be protected for much longer periods during the year.

Starlings, robins, and grackles are most destructive. Crows and cedar waxwings are particularly devastating in southern states, and at least 15 other kinds of birds can inflict some damage.

The amount of crop lost to birds varies greatly. The three largest blueberry-producing states have reported crop losses to birds ranging from 6%-20%. Losses up to 100% have been reported in some test plantings. A national loss of 10% costs growers several million dollars.

Visual or auditory repellents are sometimes used for control. The visual methods are the less effective of the two. Some common examples are hawk-like balloons, plastic strips suspended from a wire over the bushes, and artificial snakes. None of these are effective for any length of time. It has been reported that scare-eye balloons are more effective against starlings than against robins, particularly if they are moved around the planting often (Strik 1990; Brun 1990).

Auditory repellents are more effective. Many kinds of gas-operated exploders are on the market, but they are not as effective as electronic devices. This is particularly true if they are on a timer, with a constant interval between firings. Electronic devices transmit either a bird distress call or a sound that otherwise repels birds. Some of these have been quite effective in some locations. However, they can interfere with television reception and should be tried on a small scale first.

The most effective method of bird control is to cover the plantation with netting. Usually, the netting can be put on just before the berries begin to ripen and can be removed after the last harvest. It will save time and expense if the plantation has permanently installed posts and overhead wires. Posts should be at least 9 ft tall, permitting spraying and harvesting operations under the net. In larger plantations, or where over-the-row harvesters are used, netting is not feasible.

Most netting is made from synthetic materials and is treated to resist the effects of sunlight. A good-quality net, if used only during berry ripening and stored properly between seasons, can last ten years or more and is highly cost-effective. The high cost of netting should be prorated over this period. It is difficult to estimate the cost of the net, since it depends upon the type of netting, posts and wires, and the cost of labor, but it can range from $1000-$3000 per acre ($400-$1200 per ha). When purchasing netting, consider the cost, durability, ease of placement, and the amount of storage space required. Lightweight, polyethylene materials are not as durable as heavier polypropylene materials. Nylon netting is most durable.

If netting is used, permanent posts must be installed or the netting placed directly upon the bushes. The latter causes difficulty in harvesting and general maintenance, and it may not give adequate bird control. Unless the net is drawn down and tied around the base of the bush, birds can fly under it to the berries. Some fruit may also lie against the net and be exposed to birds. Further, shoots growing up through the net during the season make its removal more difficult. Suspending the net on post and wires above the plantation is much more suitable. Since posts will remain in place over a long period of time, they must be of decay-resistant material. Woods such as cedar, white oak, and black locust–preferably pressure-treated with a wood preservative–have been used, as have metal posts and posts of reinforced concrete. All of these have drawbacks, primarily expense and difficulty in getting the material. All woods rot in time, and some preservatives can injure plants. Gough and Pennucci (1981) reported on an environmentally sound, long-lasting post for use in vineyards and berry plantations. Standard 4-in (10-cm) PVC pipe filled with concrete was used to support vineyard trellises. They observed little sagging or bending, and posts showed no deterioration after ten years in the field and exposure to temperatures of 97°F to -17°F (36°C to -27°C). The cost of the posts ranged from $0.10 to $2.61 more than that of wooden posts and from $1.51 more to $1 less than metal posts, depending upon the filling material (which consisted of various concrete mixes or crushed stone). Hollow PVC pipe may be suitable for suspending lightweight netting.

OTHER ANIMAL PESTS

Growers can experience problems with two kinds of voles. This is especially a problem in Arkansas (Hancock and Draper 1989). Meadow voles are dark brown in color and 3.5-5 in (8.8-12.5 cm) long. Their bicolored tail is longer than the length of their hind legs. The animals leave small, shallow burrows and trails in high grass and gnaw the bark of plants from ground level up, removing the phloem and cambium. Plants may leaf out in spring, but leaves will remain small and the plants will die in midsummer. The pine vole is also dark brown, but has a slight reddish tinge. Body length is about 3-4.5 (7.6-11.5cm) in, and the tail is shorter than the length of its hind legs. They make extensive underground burrows and girdle roots and canes below ground. The animal lives underground. Control strategy is twofold. The first is to keep grass from growing high beneath plants by mowing or by using herbicides. This eliminates hiding places. Organic mulches can exacerbate the problem. Mow a buffer strip around the plantation to create open ground between the berry bushes and the woods, and remove nearby brush piles. The second step is baiting with poison. Place out bait in the fall, before final mowing. Mown grass will cover the bait and make it more attractive to mice. Be sure to bait around the perimeter of the plantation to intercept migrating mice. There are several compounds on the market for control of voles. Carefully follow the label instructions and your state recommendations.

Rabbits and deer also damage blueberry plants. Repellents such as tetramethylthiuramdisulfide (Thiram) can be sprayed on the plants and have been somewhat effective. A fence higher than 6 ft (3 m) is the only sure way of controlling deer effectively.

INSECTS AND MITES

There are hundreds of insects that damage blueberry plants (Phipps 1930), but only a few cause economic losses. The problem is worse in older producing areas, but it will develop with time in newer areas where blueberries have been introduced recently (Hancock and Draper 1989). Listed below are some of the more common insect pests of highbush blueberry.

Blueberry Maggot (Rhagoletis pomonella *Walsh.*)

This is probably the most serious insect pest of the blueberry. The maggot is the larva of the blueberry fruit fly and the adult looks very similar to an adult apple maggot. The Pacific Northwest does not usually suffer severe damage from this pest, and it only becomes a problem in North Carolina after very cold winters or unusually rainy harvest seasons.

Description: The adult is about the size of a housefly, but it has black bands on its wings and a pointed, black abdomen with white crossbands. Each female lays up to 100 eggs within a few weeks after emergence, depositing one egg per berry, just beneath the fruit skin. Within seven days, the eggs hatch into colorless larvae that begin feeding immediately on the fruit pulp. They turn white as they mature. After about 21 days, they drop to the ground to pupate. The puparium resembles a very small seed and is buried 1-6 (2.5-15 cm) in deep in the soil and litter. They may not emerge until the second or third summer after pupation. The peak time of emergence is just as the first berries begin to ripen.

Control: Control of the pest is difficult since protection must span the harvest season. Because emergence can be difficult to predict, a fairly sophisticated monitoring system has been developed. Saturn-yellow cardboard rectangles coated with a sticky substance trap the adult flies for monitoring populations and hatch (Prokopy and Coli 1978). Use two traps per acre. Traps may be baited with a protein hydrosylate (Meyer 1989) or ammonium acetate. Removal of wild blueberries and huckleberries will help control this pest. A good spray program is almost always necessary. Begin application when three adults per trap per week or five adults per field per week are recorded. This is usually during the early fruit maturation period.

Cranberry Fruitworm (Mineola vaccinii *Riley.*)

Like the maggot, this is a pest in most blueberry-producing regions.

Description: The adult is a small dark-grayish brown moth that generally flies at night. It emerges from the soil beneath the plants when the earliest fruit are beginning to enlarge, and the female lays her eggs inside the rim of the calyx cup. In about a week, small green caterpillars hatch and enter the berry near the stem end. They enmesh several berries in a web, then eat the fruit. Most of the interior of the berry is eaten and filled with frass. After feeding for about three weeks, the mature, green caterpillar–about 0.3 in long (9 mm)–moves to the ground to pupate and emerges the following season.

Control: Keep the plantation free from weeds and trash. This insect can be more of a problem in mulched areas and shows a preference for 'Cabot,' 'Rancocas,' 'Bluecrop,' and other earlier-ripening cultivars. Look for frass-filled, prematurely ripened berries webbed together. If you detect more than one infested cluster in five, spray immediately. In small plantings, pick and destroy infested clusters. Insecticides are applied during the first one to two post-pollination sprays.

Cherry Fruitworm (Grapholitha packardi *Zell.*)

This insect causes great economic loss in nearly all major producing areas and seriously threatens the blueberry industry.

Description: The adults are dark gray moths with deep brown bands on their wings. They fly at dusk and begin to lay greenish white, flattened eggs on the undersides of leaves during bloom, and later on green fruit. The white-bodied, black-headed larvae hatch in about a week and slowly turn pink. They enter the fruit through the calyx, and turn red a few days after beginning to feed. A single larva can destroy several berries in a cluster. About 21 days after hatching, the larvae, now about 0.25 in (6 mm) long, leave the fruit. Because entrance holes are sealed with silk, the only visible signs of infestation may be prematurely blue, shriveled berries. They overwinter in a silk cocoon, tunneled into a pruned stub or a dead twig, under bark, or in debris on the ground.

Control: Keep plantation free from trash and debris. Look for two berries webbed together to indicate their presence. Pesticides are often necessary and should be applied during the first and second post-pollination sprays.

Plum Curculio (Conotrachelus nenuphar *Herbst.*)

This is a general fruit pest that also attacks blueberries. It is a major pest in Michigan, New Jersey, and North Carolina (Hancock and Draper 1989). Early-ripening cultivars, such as 'Weymouth,' 'Earliblue,' 'Bluetta,' and 'June,' are particularly hard hit, as are plants nearest adjacent woodlands.

Description: The adult is a dark brown, 0.25-in (6-mm) long snouted beetle having light patches and four humps on its back. It overwinters as an adult in trash and organic mulch and becomes active during bloom, when it feeds on leaves and flowers. They are most active when temperatures are above 75°F (24°C). Activity is significantly reduced when temperatures fall below 70°F (21°C). The female lays a single egg in each immature green fruit, leaving a typical D-shaped puncture. After about a week, grubs hatch and feed on the inside of the fruit for about 21 days, causing them to drop prematurely. After this, the grub burrows into the soil, pupates, and, after about a month, emerges as an adult beetle, which enters hibernation in early fall.

Control: Frequent cultivation and removal of trash and debris. This pest is more prevalent on earlier cultivars. Look for the prominent egg-laying scar on the fruit and for shriveled, prematurely ripened fruit on the ground. Look for adults at dawn and dusk by placing a white sheet beneath the bushes and shaking the canes to dislodge the insects. Fallen beetles may appear dead. Apply pesticides during the first post-pollination spray if damage appears heavy.

Blueberry Blossom Weevil or Cranberry Weevil (Anthonomus musculus *Say.*)

This insect is a major pest in New Jersey and in some New England States.

Description: The adult is a small, deep-red snouted beetle that emerges during spring bud swell. It enters the clustered buds and bores small, brown holes in the unopened blossoms to feed and lay eggs. Infested flowers turn purplish, fail to open, and drop. This must not be confused with the wine color of unfertilized blossoms. The grubs eat the fallen ovary.

Control: Remove trash and wild ericacious plants from the area. Look for discolored, unopened blossoms. Apply insecticide during the delayed dormant and pre-bloom sprays.

Putnam Scale (Aspidiotus ancylus *Putnam.*)

This pest is of major concern in New Jersey, particularly in older plantings.

Description: The scales overwinter as adults, forming what appears as a dull crust on the wood surface. Crawlers emerge during bloom and move throughout the bush, sucking sap from the plant and excreting honeydew over the leaves and fruit. A black, sooty mold grows on the honeydew and interferes with photosynthesis and fruit quality. Plant vigor and yield decline. The scales congregate mostly under bark on old canes, where they appear as dull, waxy flecks. On the fruit they are encircled by red rings.

Control: Scale is a particular problem on older, neglected plantings that receive irregular pruning and on plantings hedged for machine harvesting. The pest is especially prevalent on 'Concord,' 'Bluecrop,' and 'Rancocas.' Practice good sanitation and keep bushes well pruned. It may be necessary to use superior oil as a dormant spray timed to crawler emergence.

Blueberry Bud Mite (Aceria vaccinii *Keifer.*)

This pest is a particular problem for North Carolina growers, where damage is sporadic and highly correlated to weather, with mild winters contributing to more severe infestations.

Description: In spring, mites leave infested bud scales and floral parts and migrate to areas beneath the scales of newly forming buds. There, they mate, feed, and deposit eggs. As the season progresses, they burrow toward the center of the bud. Feeding and egg-laying continue into the winter. Injured plant parts show a roughened, red discoloration. Sometimes, small red pimples appear on fruit. Buds fail to expand and produce fruit.

Control: This pest is often more of a problem on early cultivars such as 'Weymouth,' 'Rancocas,' 'Pioneer,' and 'Scammell.' It is also becoming a problem on 'Berkeley.' 'Bluechip' appears resistant. Look for roughened bud parts that display red discoloration. Post-harvest application of insecticides or miticides with high-pressure nozzles may be necessary.

Black Vine Borer (Brachyrhynus sulcatus *Fabricius.*)

This insect causes substantial damage in the Pacific Northwest, where it also attacks strawberries.

Description: Grubs hibernate in the bush crown, while adults overwinter in debris beneath the plants. They emerge as adult weevils in early summer. The black adults are about 0.5 in (1.25 cm) long, with gold and white patches on their wings. The adults, which are all female, reproduce without mating and begin laying eggs as berries begin to ripen. Eggs are deposited in soil near the plant crown, where the feeding grubs can girdle the plant just below the soil line. Affected plants are weakened, and their leaves redden prematurely.

Control: Rogue out severely infested plants or prune out affected areas.

Blueberry Stem Borer (Oberea *sp.*)

This is a serious pest common in North Carolina.

Description: Adults are light brown beetles with long antennae and are active in early to mid summer. The female lays her eggs just under the bark near the tip of new shoots. These are deposited between two parallel rings of punctures cut around the stem about 4 in (10 cm) from the tip. The shoot tip wilts and blackens. Yellowish larvae tunnel upward, killing the tip of the shoot, then bore downward 2-10 in (5-25 cm) during the first season. Frass is ejected through small holes chewed to the outside. Larvae resume feeding in the spring, until they reach the base of the stem, where they overwinter a second time. By the third season, the larvae have attacked several stems and damage is considerable. They then overwinter again, pupate the following spring, and emerge as adults. Canes can be completely killed.

Control: Remove infested tips (by cutting below the tunneled portion) and infested canes. Look for mounds of loose frass near the base of plants and for canes with reddened leaves, or defoliated canes, having frass-filled pin holes. Remove wild blueberry, huckleberry, rhododendron, and laurel, since these also are host plants for this insect.

Root Weevils

Root weevils include five especially troublesome species: the strawberry root weevil (*Otiorhynchus ovatus*); the rough strawberry root weevil (*Otiorhynchus rugosotriatus*); the black vine weevil (*Otiorhynchus sulcatus*); the woods weevil (*Nemocestes incomptus*); and the obscure root weevil (*Sciopithes obscurus*).

All five of these weevils can cause significant economic damage to blueberries in the Pacific Northwest, though the first three are particular problems.

Description: The weevils are usually black, brown, or gray in color. The black vine weevil has orange spots on its back. Larvae have brown heads and white bodies usually bent into a "C" shape. They overwinter while feeding on roots, pupate in mid spring, and hatch into adults late in the bloom season. Adults attack the leaves, usually at night, return to soil during day, and lay eggs in midsummer. Larvae hatch and burrow into the soil to begin feeding on roots again.

Control: Woodlands and fence rows harbor the insects. Precise identification is important since not all insecticides are equally effective against all species. Look for notching of the leaves to indicate feeding adults. Pre-plant soil fumigation will control these weevils. Also, the obscure root weevil can be controlled with some post-bloom sprays.

Sharpnosed Leafhopper (Scaphytopius magdalensis Provancher.)

This insect is of concern not so much for the foliage damage it causes but because it transmits the stunt pathogen.

Description: The wedge-shaped, brown-gray adults are about 0.18 in (5 mm) long and appear in midsummer. They lay eggs in leaves shortly after their appearance. The nymph is distinguishable by a white, hourglass-shaped mark on its back.

Control: Practice good sanitation and an adequate spray program. Usually, first-generation control is accomplished with sprays for fruitworm and curculio.

Other Insects

Other insects, such as adult Japanese beetles, can destroy foliage and fruit, while their grubs damage blueberry roots, especially under organic mulches. Leaf tiers, leaf rollers, leaf miners, sawfly, yellow-necked caterpillar, acronicta caterpillar, fall webworm, and *Prionus* borers can all cause some damage. The eastern subterranean termite (*Reticulitemes flavipes* Kollar) occasionally feeds on live blueberry roots. The gypsy moth can be a problem in the east, where oak trees are common. These often arrive about the time of bloom, making spraying difficult. All of these pests can be controlled with proper plantation management and close attention to pesticide controls.

DISEASES

Scores of pathogens–fungal, bacterial, viral, and others–attack the highbush blueberry. Several of these cause serious losses to the crop and, in some cases, to the plants themselves. As with insect pests, problems with diseases are often worse in older producing areas. The more important diseases are described below.

Stem Canker (*Botryosphaeria corticis, Fusicoccum putrefaciens, Cornyeum microstictum, Phomopsis vaccinii, Botrytis cinerea*)

All of these fungi cause cankers on twigs and stems. Each will be discussed briefly. 'Bluechip' is resistant to all canker in North Carolina. The use of clean cutting wood is paramount to control of cankers (Cline 1990).

Botryosphaeria corticis (Demaree and Wilcox) Arx and Muller

Description: This fungus is found in most blueberry-producing areas, though the southern United States has been particularly hard hit. The pathogen invades new shoots through lenticels on the sunny sides of the shoot, and red swellings appear at that point by early fall. Black fruiting bodies later develop on these swellings. During the second growing season, the red coloration disappears and fissures develop at points of infection. These turn gray. In following years, the gray canker completely girdles the cane. Black fruiting bodies on the cankers release spores during rainy periods. These infect new shoots and the cycle begins anew.

Control: Rogue all infected plants and cut out all cankered canes. 'Atlantic,' 'Jersey,' and 'Scammell' are highly resistant, while 'Pemberton,' 'Rancocas' and 'Rubel' show some resistance. 'Cabot,' 'Dixi,' 'Concord,' 'June,' 'Pioneer,' 'Stanley,' and 'Weymouth' are extremely susceptible.

Fusicoccum putrefaciens Shear (Godronia cassandrae Peck)

This fungus can be a problem in northern producing areas, particularly Michigan, parts of Canada, and Europe.

Description: The life cycle of this pathogen is similar to that of *Botryosphaeria corticis,* with infection often taking place through a leaf scar. The enlarging canker turns gray, but can retain red margins. Infected canes wilt suddenly under summer stress.

Control: Same as for other cankers. 'Blueray,' 'Rancocas,' 'Stanley,' 'Berkeley,' and 'Rubel' show some resistance, while 'Jersey' and 'Earliblue' are highly susceptible.

Cornyeum microstictum Berk. and Br.

This can sometimes become a problem in the northeastern United States.

Description: The pathogen generally enters the shoot through a wound, resulting in a canker that eventually girdles the cane.

Control: Rogueing and careful pruning.

Phomopsis vaccinii

This disease can be a problem in the major producing areas of Michigan, New Jersey, and North Carolina. It also appears in New England, attacking both canes and twigs. It more commonly causes problems on weakened or wounded bushes.

Description: Young shoots are invaded first, primarily through flower buds at budbreak (Milholland 1982). The fungus moves steadily into older tissue, canes, and, finally, the crown. Red spots sometimes appear on leaves of infected shoots. Because shoot tips blacken in the twig-blight stage, phomopsis is often taken at first glance as frost or winter injury. The fungus also attacks the fruit.

Control: Remove all infected tissue and maintain plants in vigorous condition. Plants of 'Harrison,' 'Croatan,' 'Blueray,' and 'Earliblue' are highly susceptible to infection, while those of '1613-A,' 'Rancocas,' and the German cultivar 'Heerman' show some resistance. Fungicide sprays at mid-bloom may be helpful.

Botrytis cinerea (Gray Mold)

This is found throughout blueberry-producing areas, but it is considered a major problem in the southeastern United States, New Jersey, and the Pacific Northwest (Hancock and Draper 1989). It is also a problem in the Canadian Maritimes.

Description: The pathogen overwinters in previously infected plant material. In spring, spores are carried by rain and wind to new tissue. Cool, wet periods for up to a week can result in serious infection. Infected tips of new shoots turn black and finally gray. Infected corollas turn brown, display abundant mycelia, and remain attached to the ovary. Eventually, they fall and can stick to newly emerged leaves, spreading the pathogen. *Botrytis* can result in considerable fruit spoilage after harvest.

Control: Avoid overfertilization, since rapid succulent growth is more susceptible to attack. Poor pollination and overhead irrigation during bloom both increase incidence of this disease. Follow a thorough fungicide spray program, especially during mid-bloom and the first post-pollination sprays. *Botrytis* in the Pacific Northwest and in Florida is mostly resistant to benomyl.

Powdery Mildew (Microsphaera penicillata var. vaccinii (Schw.) Cooke)

This is considered a major disease in the southeastern United States, and it is widespread in the Michigan area. Loss occurs through gradual weakening of the plants.

Description: The pathogen attacks the leaves, covering the upper or lower surfaces, or both, with white mycelium. Usually the lower leaf surface is attacked first. Red-bordered chlorotic areas appear on the upper surface. Directly beneath these, water-soaked areas will appear on the lower surface later in the season. By the end of the season, orange (and later black) fruiting bodies develop on the lower leaf surface.

Control: Practice good sanitation in the planting and prune for good air circulation. 'Jersey' is very susceptible.

Stem Blight (*Botryosphaeria dothidea* (Moug. ex Fr.) Ces. and deNot.)

This causes major concern in North Carolina and Arkansas, and is a major cause of failure of one and two-year old plantings (Cline and Milholland 1992). It is often confused with winter damage and cane canker. 'Bluechip' is especially susceptible.

Description: This pathogen commonly attacks only one side of a stem. Infected wood turns brown and can extend the entire length of the cane. During the initial stage of infection, leaves can become chlorotic or red. The entire branch usually dies, while other branches close by remain unaffected.

Control: Keep plants vigorous. Prune out affected branches immediately, well below visibly infected tissue. Because the pathogen often enters through pruning wounds, avoid pruning until the fourth year after planting. Where feasible, treat pruning cuts with a wound dressing immediately. Cline and Milholland (1992) reported that dipping blueberry roots into a benomyl-kaolin slurry gave three to five months protection against development of stem blight in container-grown nursery plants.

Phytophthora Root Rot (*Phytophthora cinnamoni* Rands)

This disease is often associated with wet areas and is of particular concern in Arkansas and North Carolina. It is the primary cause of plant death in Florida.

Description: This fungus destroys the fibrous roots of the blueberry. Above-ground symptoms include chlorosis and reddening of the leaves, early defoliation, reduction in shoot growth, and vascular discoloration in both crowns and canes. Wilted branches are particularly noticeable in early morning. Near the ground level, brown canes and red bands can be seen extending up the vascular tissue.

Control: Do not plant in poorly drained soils. Plant on raised beds (1.5 ft or 0.5 m high and 5 ft or 1.7 m wide) with sloping sides. Do not overirrigate and do maintain soil pH below 5.5. Soil drenched with some fungicides may be effective in some areas. 'Patriot' is resistant.

Anthracnose (Glomerella cingulata [Spaulding and von Schrenk])

This is a problem in the southeastern United States, New Jersey, and Michigan.

Description: The fungus overwinters in previously infected dead tissue and is spread by warm spring rains. It infects the young fruit, then remains dormant until the fruit mature, at which time the fruit surface, particularly at the blossom end, will pucker and become covered with pinkish spores. During warm, humid springs, the fungus can also cause stem cankers, blossom cluster blight, and leaf spots. The latter vary from small spots to large, dead areas. The earliest symptom of infection is shoot blight.

Control: Remove infected tissue where possible. 'Jersey,' 'Blueray,' and 'Harrison' are very susceptible.

Mummyberry (Monilinia vaccinii-corymbosi [Reade] Honey)

This fungus is very widespread and affects plants in nearly every geographic area. It is the most common blueberry disease in the Pacific Northwest (Doughty et al. 1981). It also accounted for a loss of nearly 10% of the crop in North Carolina in 1989 (Cline 1990).

Description: The fungus overwinters in small gray globes on the ground. Cup-like structures form with spring rains and these discharge spores during rainy periods at about the time of bud swell. Look for cups in the transition zone between wet row middles and dry row ridges. The spores infect only very young growth of shoots and flowers, which blacken and wilt suddenly. Fruiting bodies appear on the dead tissue and are carried to open flowers by bees and wind. Infection of the ovaries occurs, and developing seeds abort. Fruit appear to develop normally nearly to maturity, but instead of turning blue, infected fruit develop a salmon color and fall to the ground. The flesh is replaced by fungal tissue and the skin falls away, leaving the mummy.

Control: Late-fall or early-spring disking will bury the mummies before the infection period begins. Applying at least 2 in (5 cm) of organic mulch at this time can also help bury mummies. A mixture of equal parts urea and sand–spread beneath bushes at the rate of 200 lb per acre (224 kg per ha) during spring bud break–has been effective, though it can stimulate early growth. Apply the urea after cultivation since it is not as effective in controlling the crops if it is incorporated. Apply delayed dormant and mid-bloom sprays. 'Spartan,' 'Atlantic,' and 'Collins' are partially resistant to both phases of the disease. 'Bounty,' 'Croatan,' 'Jersey,' and 'Morrow' are highly susceptible.

Stunt

This is a major disease of blueberries in most regions. Once thought to be caused by a virus, it is now known that the pathogen is a mycoplasma.

Description: Symptoms vary with cultivar, season, and growth stage. Generally, leaves develop interveinal and marginal chlorosis, and become cupped, puckered, and smaller than normal. The chlorotic areas turn a bright red in late summer, before the onset of normal fall coloration. The red color develops in two strips near the leaf margin, parallel to the midrib. Usually, leaves on the upper portion of the shoot show symptoms. Symptoms of magnesium deficiency, which are similar, occur on the lower leaves. Berries are small, dark, and bitter and remain attached to the plant. Branches are twiggy and bushy from the breaking of normally dormant buds along the canes.

Control: The disease is more severe in heavily pruned plants. It is transmitted by the sharpnosed leafhopper, and controlling this insect is of great importance in controlling stunt. The pathogen cannot be transmitted by pruning equipment. Rogue infected plants promptly and completely. Treat infected plants with a suitable insecticide to kill leafhoppers before digging them out. This will reduce the chances of leafhoppers moving to other plants during the removal process.

Shoestring

This is a viral disease of major concern in Michigan.

Description: Plants, most often those that are heavily pruned, develop a red discoloration of the leaf midrib. These leaves become narrow, wavy, and crescent-shaped (or strap-like) and can develop further red coloration. Young shoots develop red streaks of varying length along the sides exposed to sunlight. Shoots become twisted, spindly, and brittle. Blossoms may be streaked with red, and immature green berries can have a purplish hue on the exposed side.

Control: Rogue diseased plants. Follow a careful insect control program, since the disease is spread by aphids (*Illinoia pepperi*).

Red Ringspot

This viral disease is becoming common in northern blueberry areas of New Jersey and Michigan.

Description: Red spots become noticeable on the upper surface of leaves as the growing season advances. Spots caused by powdery mildew are not as well defined and they develop on both surfaces. Older leaves near the center and base of the bush show symptoms first. Red spots or rings can also develop on stems and fruit. Rings and spots on 'Blueray' are chlorotic instead of red.

Control: Rogue diseased plants. The vector that spreads this disease is unknown.

Bacterial Canker (Pseudomonas syringae Van Hall)

This disease is of major economic importance in the Pacific Northwest.

Description: In winter, water-soaked areas appear on year-old canes. These turn brown or black and can extend the entire length of the stem. All buds in the infected areas die. Infection often occurs through frost-injured tissue and can be confused with *Botrytis*, scorch disease, and frost damage.

Control: Prune out affected stems. Copper sprays during the formation period may be effective. 'Rancocas' and 'Weymouth' are highly resistant.

Other Diseases

Other diseases of the highbush blueberry include crown gall, root gall, mosaic, necrotic ringspot, armillaria root rot, several kinds of leaf spots, rusts, witches' broom, and leaf mottle. These can become troublesome under certain conditions. The leaf-mottle virus is transmitted by honeybees and infected plants should be rogued and destroyed (Boylan-Pett et al. 1990).

Nematodes

These microscopic, worm-like animals can cause significant damage in some areas under certain conditions. Several species can cause some damage to root systems, particularly in cutting beds. *Tetylenchus christiei*, the stubby root nematode, attacks new roots that are emerging from callus tissue, resulting in poorly rooted, or dead, cuttings. *Xiphinema americanum* Cobb., the dagger nematode, is the vector for tobacco ringspot virus, which causes necrotic ringspot in blueberry.

Control: Control is almost impossible in established plantings. New areas known to be infested with these species can be fumigated before planting. Propagating beds should be sanitary, and only steamed or fumigated soil should be used.

GENERAL PEST CONTROL TACTICS

Control of many pests can be enhanced by practicing good sanitation around the plantation. Remove infested plant material, weeds, and prunings as they develop. Brush piles should be cleared from the area, and wild ericaceous plants or other host plants should be destroyed. No matter how tidy the plantation, chemical control is sometimes necessary. Be sure you have adequate training in applying pesticides. To reduce drift to non-target organisms and areas, do not spray when wind speed is greater than 5 mph. Use low nozzle pressure and a large-orifice nozzle. Spray as close to bushes as possible. Also, spray when relative humidity is between 50% and 80%, since lower humidity causes rapid evaporation of spray droplets and decreases spray volume. Spreader stickers can also be used to improve distribution of spray on plant surfaces. Appropriate warnings of pesticide application should be given to workers and the fields should be posted with worker re-entry periods. No one should re-enter treated fields without protective clothing, at least until spray material has dried.

Always follow all safety precautions on the material label, and keep all pesticides locked away until use.

PROTECTING HONEYBEES

Since your crop depends upon the direct action of bees, take all precautions to ensure that they are not injured by pesticides (Table 15-1). Most damage occurs when insecticides are applied improperly during bloom or when adjacent flowering crops and water are contaminated by drift. Insecticide dusts collected with pollen and returned to the hive are particularly dangerous. Never spray when bees are in the plantation and always notify local beekeepers of your intent to spray. Compounds that are highly toxic to bees should be applied during cooler weather and at night, when bees are not active. Do not spray when heavy dews or unusually cool temperatures are expected, since both will delay drying time. Further, unusually high temperatures may cause bees to forage at normally inactive times. If several compounds give effective control, always use the least toxic.

Commonly, symptoms of bee poisoning include a great number of dead bees outside the hive; regurgitation and partial paralysis (Lindane poisoning); and dazed activity (chlorinated hydrocarbons). Carbaryl poisoning causes lethargy, and death may come in two to three days. Dead brood near the hive entrance is symptomatic of Carbaryl and arsenical poisoning. Queens may be killed by these slower-acting compounds, severely weakening the colony.

Because pesticide recommendations vary among locations and change very rapidly, any spray schedule formulated here would be obsolete before this book was printed. Therefore, always consult with your local Cooperative Extension Service or state university for the most recent recommendations, and carefully follow all label instructions.

Table 15-1. Toxicity of some pesticides to bees.

Highly Toxic	Moderately Toxic	Slightly Toxic
Guthion	Thiodan	Lime Sulfur
Parathion		Funginex
Imidan		Benlate
Lannate		Dipel
Sevin		Javelin
Diazinon		Captan
Malathion		

Chapter 16

Marketing

In retail marketing, the grower provides the harvesting and packaging at some considerable cost, which is passed along to consumers through a roadside stand or farm market. Wholesale marketing allows the grower to market more fruit, but at a lower price. Growers should first consider their location, market potential, and competition before deciding which sales approach to take. Marketing decisions should be made even before planting begins. Adequate parking and safe movement of customers need to be planned early for on-farm sales.

WHOLESALE

The grower producing for the wholesale market must consider the supply and reputation of the businesses who will do the marketing. Local grocery stores and convenience stores can be a good market for blueberries. Finer restaurants may be willing to accept top-quality berries for desserts made on the premises. Chain stores sometimes purchase locally grown produce, but this market may be difficult to enter because such stores often purchase in huge bulk quantities and already have their supply network in place.

In recent years, wholesale farmers markets have increased throughout the country. Operating in these can take considerable time, and you must be certain that you will have enough berries for sale through the season. Nearby processing plants offer very large growers another marketing outlet. Whatever wholesale markets are used, the grower must always confer with the buyer on what quality, quantity, and delivery times are expected. The grower also needs to consider the purchase of handling and grading equipment, standard-sized containers, and pallets.

RETAIL

Retail, or direct marketing, provides berries directly to the consumer. This is accomplished through the establishment of roadside stands, individual booths at farmers markets, and U-Pick operations.

Recent marketing trends point toward development of conveniently located roadside stands where customers can purchase fresh, neatly packaged, high-quality fruit. Most of these offer more than one product for sale. Vegetables, other fruit such as raspberries and peaches, and eggs and dairy products are often found next to fresh blueberries. Some stands even complement fresh sales with homemade or locally made preserves and pies. This can be especially attractive if they are made from blueberries, and their recipes are provided free. Most customers visiting these stores are neighbors who live within 10-20 miles of the farm. They can be very demanding, expecting high-quality fruit arranged on full, orderly display shelves. These shelves are often refrigerated to prolong freshness. Inferior-quality fruit should be carefully separated and used in preserves and baked goods. Workers at retail stands must be organized, neat, and helpful. A good knowledge of the product can often make a sale.

Individual booths at local farmers markets provide another outlet for sales. Again, a neat appearance, friendly workers, and the highest-quality fruit are necessary for high sales volume. Arrangements for marketing in such places are made well ahead of time, and the grower must be sure he will have enough produce through the entire season. Prompt arrival, set-up, and take-down times must be carefully observed.

U-Pick operations, where the consumer actually harvests the fruit, are the principal market for direct sales. Often this means the highest-quality fruit at a reasonable price. It provides a family day together at the farm and allows consumers to purchase in a volume that many could not afford otherwise. The grower reduces harvesting and packing problems and no longer must depend upon the often unreliable supply of hired pickers. However, many fruit may not be picked, and inclement weather will affect attitudes toward picking. Growers must also hire salespeople and field supervisors, which reduce profits. Consumers can be encouraged to provide their own containers, which are weighed-in upon arrival, or the grower can supply these. By supplying uniform containers, the grower eliminates the need for scales and the time required for weighing the produce at check-out. Many customers are not accustomed to rural life. Excessive dust, mud, and weeds and poor roads or directions are turn-offs for them. Parking areas should be handy, well-designated, and designed for efficient entrance and exit. Sixty-degree-angle parking requires about 1000 sq ft for every 20 cars, while ninety-degree parking accommodates 30 cars in the same space. Parking some distance from the harvest area may make transportation on a farm wagon attractive, but it will also increase liability insurance. Vigorous promotion amounts to about 3%-5% of the total budget for retail sales. Coverage in local newspapers or on radio stations and posters well placed in local stores are all effective. Always advertise and adhere to a regular picking schedule, which should rarely extend beyond 3 p.m. Most customers of such operations live within a 15-mile radius of the plantation, so it makes little sense to advertise for more distant markets. The average customer harvests about 12 lb (5.5 kg) of fruit. An average per-acre yield of 6000 lb (6720 kg per ha) therefore requires 500 sales per acre. Since only about 10% of the population will visit a U-Pick operation, each acre of berries requires a population density of 5000 people within a 15-mile radius. The heaviest harvest pressure is on the weekends, so advertise accordingly.

Having customers sign a guest book provides a simple way to get addresses for postcard mailings announcing the expected beginning of harvest the following year. Often, over 50% of the customers are repeats. Remember, a satisfied customer will tell two to three people of his experience, but an unsatisfied one will tell eight. Word of mouth is the best and least expensive advertisement. To promote a satisfying experience, provide adequate and polite "people control." Locate several experienced field supervisors throughout the area and equip them with bullhorns. Post rules at the entrance and strictly, but politely, enforce them. Do not hesitate to remove disobedient customers from your fields. Allow only one exit and one entrance from the area, both of which should pass directly by the front stand. Mark harvest areas well, and restrict the public to those areas. Harvesting immature fruit will only turn the customer sour on your operation and decrease the volume and profit of the crop. Provide tight security for your cash receipts and bank them promptly. Instruct your workers to be polite and helpful. Encourage them to enter into brief conversations with customers and to aid them in identifying and harvesting ripe fruit. If customers do not harvest all the ripe fruit from a bush, supervisors can indicate politely that many ripe fruit were "missed," and may even harvest some of them.

Some growers restrict small children from entering the picking area. This should be posted clearly at the entrance to the field. So as not to deter potential customers, growers might provide playground equipment and babysitters next to the picking area. Often, picnic areas encourage customers to make a day of it, which can increase sales substantially.

A check-in stand should be located near the field entrance. Upon arrival, the customer will check in at the stand where a container of known weight is provided. Upon leaving, the customer will check out and the cost of the harvest determined. Be sure that the scale is adjusted for the weight of the empty container and that the container cost is figured into the total cost of the fruit. If the customer provides his own container, its weight should be recorded upon arrival. Volume sales (such as by the quart or liter) can be difficult, since by heaping the container, the customer can obtain 10%-20% more fruit at no extra cost. If such sales are used, the price should be increased by 20%, or sales should be made in terms of "struck" containers only.

Refreshments and restroom facilities can be provided at the checkout station to further increase the customers' comfort and satisfaction. Some growers also provide free recipes and information on blueberry culture to inquisitive customers.

Lastly, be sure no hazardous conditions exist in your fields. Unused farm equipment should be moved to a safe place, irrigation lines clearly marked, and ruts and animal holes filled. Always check with your insurance agent to be sure you have enough liability coverage.

A modification of the traditional U-Pick operation is to have local church, scouting, or 4-H groups pick fruit at the peak of harvest. The growers can offer such groups special prices, and if they bring along their own supervisors, the regular field supervisors can have the day off. U-Pick can also be combined with wholesale methods during bad weather conditions, poor public turnout, and excessive production. (Funt, personal communication).

References

Abbott, J.D. and R.E. Gough. 1986. Split-root water application to highbush blueberry plants. *HortScience* 21(4):997-998.

_____. 1987a. Seasonal development of highbush blueberry roots under sawdust mulch. *Journal of the American Society for Horticultural Science* 112(1):60-62.

_____. 1987b. Growth and survival of the highbush blueberry in response to root zone flooding. *Journal of the American Society for Horticultural Science* 112(4):603-608.

_____. 1987c. Prolonged flooding effects on anatomy of highbush blueberry. *HortScience* 22(4):622-625.

_____. 1987d. Reproductive response of the highbush blueberry to root-zone flooding. *HortScience* 22(1):40-42.

Ascher, S. 1991. Just Between Us. The Blueberry Bulletin. Rutgers University Cooperative Extension Service VII (17):2.

Anonymous. 1991. "Sunrise" Highbush Blueberry. *Northland Berry News* 5(3):90.

Arnold, J.T. and C.F. Thompson. 1982. Chlorosis in blueberries: A soil-plant investigation. *Journal of Plant Nutrition* 5:747-753.

Austin, M.E. and K. Bondari. 1992a. Soil pH effects on yield and fruit size of two rabbiteye blueberry cultivars. *Journal of Horticulture Science* 67(6):779-785.

_____. 1992b. Hydrogel as a field medium amendment for blueberry plants. *HortScience* 27(9):973-974.

Austin, M.E. and A.D. Draper. 1987. 'Georgiagem' blueberry. *HortScience* 22(4):682-683.

Bachelard, E.P. and F. Wightman. 1974. Biochemical and physiological studies on dormancy release in tree buds. III. Changes in endogenous growth substances and a possible mechanism of dormancy release in overwintering vegetative buds of *Populus balsamifera. Canadian Journal of Botany* 51:2315-2326.

Bailey, J.S. 1938. The pollination of the cultivated blueberry. *Pro-*

ceedings of the American Society for Horticultural Science 35:71-72.

Bailey, J.S. and L.H. Jones. 1941. The effect of soil temperature on the growth of cultivated blueberry bushes. *Proceedings of the American Society for Horticultural Science* 38:462-464.

Bailey, J.S. and J.L. Kelley. 1959. Blueberry growing. University of Massachusetts College of Agriculture Publication Number 240.

Bailey, J.S., H.J. Franklin, and J.L. Kelley. 1939. Blueberry culture in Massachusetts. Massachusetts Agricultural Experiment Station Bulletin 358.

Baker, R.E. and H.M. Butterfield. 1951. Commercial bushberry growing in California. California Agricultural Extension Service Circular 169.

Ballinger, W.E., A.L. Kenworthy, H.K. Bell, E.J. Benne, and S.T. Bass. 1958. Production in Michigan blueberry plantations in relation to nutrient element content of the fruit-shoot leaves and soil. Michigan Agricultural Experiment Station Quarterly Bulletin 40:896-905.

Ballinger, W.E., E.P. Maness, and L.J. Kushman. 1970. Anthocyanins in ripe fruit of the highbush blueberry, *Vaccinium corymbosum* L. *Journal of the American Society for Horticultural Science* 95:283-285.

Ballington, J.R. 1989. Variety selection and breeding. Proceedings of the 23rd Annual Open House, Southeastern Blueberry Council. North Carolina State University: Elizabethtown, North Carolina. pp. 16-17.

Ballington, J.R., B.W. Foushee, and F. Williams-Rutkosky. 1989a. Potential of chip-budding, stub grafting or hot callusing following saddle-grafting on the production of grafted blueberry plants. *Proceedings of the Sixth North American Blueberry Research and Extension Workers Conference.* Portland, Oregon. pp. 114-120.

Ballington, J.R., C.M. Mainland, S.D. Duke, A.D. Draper, and G.J. Galletta. 1989b. 'Bounty' highbush blueberry. *HortScience* 24(1):161-162.

Ballington, J.R., C.M. Mainland, S.D. Rooks, A.D. Draper, and G.J. Galletta. 1990a. 'Blue Ridge' and 'Cape Fear' southern highbush blueberries. *HortScience* 25(12):1668-1670.

Ballington, J.R., S.D. Rooks, and C.M. Mainland. 1990b. 'Reveille' a new southern highbush blueberry cultivar for mechanical harvesting for both fresh and processing market outlets. *Proceedings of the 24th Annual Open House, Southeastern Blueberry Council.* North Carolina State University, Clinton, North Carolina.

Beaudry, R.M., A.C. Cameron, A. Sherazi, and D.L. Dostal-Lange. 1992. Modified-atmosphere packaging of blueberry fruit: Effect of temperature on package O_2 and CO_2. *Journal of the American Society for Horticulture Science 117*(3):436-441.

Beckwith, C.S. 1930. Report of the Department of Entomology. New Jersey Agricultural Experiment Station, New Brunswick, New Jersey. p.174.

Benson, A.B. 1966. *The America of 1750: Peter Kalm's travels in North America.* New York, NY: Dover Publications.

Blasing, D. 1989a. A review of *Vaccinium* research and the *Vaccinium* industry of the Federal Republic of Germany. *Acta Horticulturae* 241:101-109.

_____ . 1989b. Performance of highbush blueberries on sites previously used for agricultural crops. *Acta Horticulturae* 241:213-220.

Boller, C.A. 1956. Growing blueberries in Oregon. Oregon Agricultural Experiment Station Bulletin 499 (rev.).

Boylan-Pett, W., D. Ramsdell, and R. Hoopingarner. 1990. In hive transfer, pollen exchange, longevity of blueberry leaf mottle virus (BBLMV) in pollen, and bee transmission of BBLMV to blueberries. *Proceedings of the Sixth North American Blueberry Research and Extension Workers Conference.* Portland, Oregon. pp. 140-145.

Bradley, R., A.J. Burt, and D.J. Read. 1982. The biology of mycorrhiza in the Ericaceae. VIII. The role of mycorrhizal infection in heavy metal resistance. *New Phytologist* 91:197-209.

Brightwell, W.T. 1941. Yield, size of berries, and season of maturity of the highbush blueberry as influenced by severity of pruning. *Proceedings of the American Society for Horticultural Science* 38:447-450.

Brightwell, W.T. and S. Johnston. 1944. Pruning the highbush blueberry. Michigan Technical Bulletin 192.

Brown, J.C. and A.D. Draper. 1980. Differential response of blueberry (*Vaccinium*) progenies to pH and subsequent use of iron.

Journal of the American Society for Horticultural Science 105:20-24.

Brun, C.A. 1990. Overview of the Pacific Northwest blueberry industry. *Proceedings of the Sixth North American Blueberry Research and Extension Workers Conference.* Portland, Oregon. pp. 6-12.

Burton, C.L., B.R. Tennes, and G.K. Brown. 1979. Postharvest hot-water and fungicide treatments for reduction of decay of blueberry. *Proceedings of the Fourth North American Blueberry Research Workers Conference,* University of Arkansas, Fayetteville. pp. 141-158.

Cain, J.C. 1952. A comparison of ammonium and nitrate N for blueberries. *Proceedings of the American Society for Horticultural Science* 59:161-166.

_____. 1954. Blueberry chlorosis in relation to leaf pH and mineral composition. *Proceedings of the American Society for Horticultural Science* 64:61-70.

Camp, W.H. 1945. The North American blueberries with notes on the other groups of Vacciniaceae. *Brittonia* 5:203-275.

Cappellini, R.A., A.W. Stretch, and J.M. Mariello. 1972. Fungi associated with blueberries held at various storage times and temperatures. *Phytopathology* 62:68-69.

Card, F.W. 1903. Bush Fruits. Rhode Island College of Agriculture and Mechanical Arts Bulletin 91.

Carlson, J.D. and J.F. Hancock. 1991. A methodology for determining suitable heat-unit requirements for harvest of highbush blueberry. *Journal of the American Society for Horticultural Science* 116(5):774-779.

Christopher, E.P. and V. Shutak. 1947. Influence of several soil management practices upon yield of cultivated blueberries. *Proceedings of the American Society for Horticultural Science* 49:211-212.

Clark, J.R. and J.N. Moore. 1991. Southern highbush blueberry response to mulch. *HortTechnology* 1(1):52-54.

Clayton-Greene, K. 1989. The blueberry industry in Australia: an overview. *Acta Horticulturae* 241:91-93.

Cline, W.O. 1990. Disease control in 1990 with limited supplies of Difolitan. *Proceedings of the 24th Annual Open House of the*

Southeastern Blueberry Council. North Carolina State University, Clinton, NC.

Cline, W.O. and R.D. Milholland. 1992. Root dip treatments for controlling blueberry stem blight caused by *Botryosphaeria dothidea* in container-grown nursery plants. *Plant Disease* 76(2):136-138.

Collison, R.C. 1942. Making soils acid for blueberries. Farm Research New York Agricultural Experiment Station 8:18.

Coorts, G.D. and J.W. Hull. 1972. Propagation of highbush blueberry (*Vaccinium corymbosum*) by hard- and softwood cuttings. *Plant Propagator* 18:9-12.

Corliss, J. 1992. What's a better blueberry pollinator? *Agricultural Research*, March, p. 19.

Costante, J.F. and B.R. Boyce. 1968. Low temperature injury of highbush blueberry shoots at various times of the year. *Proceedings of the American Society for Horticultural Science* 93:267-272.

Coville, F.V. 1910. Experiments in Blueberry Culture. United States Department of Agriculture Bureau of Plant Industry Bulletin 193.

_____. 1921. Directions for blueberry culture. United States Department of Agriculture Bulletin 974.

_____. 1937. Improving the wild blueberry. United States Department of Agriculture Yearbook.

Crane, J.H. and F.S. Davies. 1987. Hydraulic conductivity, root electrolyte leakage, and stomatal conductance of flooded and unflooded rabbiteye blueberry plants. *HortScience* 22:1249-1252.

_____. 1988. Flooding duration and seasonal effects on growth and development of young rabbiteye blueberry plants. *Journal of the American Society for Horticultural Science* 113:180-184.

_____. 1989. Flooding responses of *Vaccinium* species. *HortScience* 24(2):203-210.

Cronon, W. 1983. *Changes in the Land*. New York, NY: Hill and Wang.

Cummings, G.A. 1989. Fertilization. *Proceedings of the 23rd Annual Open House of the Southeastern Blueberry Council*. North Carolina State University. Elizabethtown, North Carolina. pp. 33-34.

Cummings, G.A., C.M. Mainland, and J.P. Lilly. 1981. Influence of soil pH, sulfur, and sawdust on rabbiteye blueberry survival, growth, and yield. *Journal of the American Society for Horticultural Science* 106:783-785.

Dale, A., R.A. Cline, and C.L. Ricketson. 1989. Soil management and irrigation studies with highbush blueberries. *Acta Horticulturae* 241:120-125.

Darlington, W. 1847. *Agricultural Botany.* Philadelphia, PA: J.W. Moore.

Darrow, G.M., J.H. Clark, and E.B. Meader. 1939. The inheritance of certain characters in the cultivated blueberry. *Proceedings of the American Society for Horticultural Science* 37:611-616.

Davies, F.S. and C.R. Johnson. 1982. Water stress, growth, and critical water potentials of rabbiteye blueberry (*Vaccinium ashei* Reade). *Journal of the American Society for Horticultural Science* 107:6-8.

Davies, F.S. and J.A. Flore. 1987. Gas exchange and flooding stress of highbush and rabbiteye blueberries. *Journal of the American Society for Horticultural Science* 111:565-571.

Dijkstra, J. and J.M. Wijsmuller. 1989. Experiments with blueberries in the Netherlands. *Acta Horticulturae* 241:87-90.

Doehlert, C.A. 1937. Blueberry tillage problems and a new harrow. New Jersey Agricultural Experiment Station Bulletin 625.

Doughty, C.C., E.B. Adams, and L.W. Martin. 1981. Highbush Blueberry Production. Pacific Northwest Extension Bulletin 215.

Douglas, J. 1967. The propagation of highbush blueberries by softwood cuttings. International Society for Horticultural Science Working Group Symposium on Blueberry Culture in Europe. Venlo, Netherlands. pp. 95-104.

Draper, A.D., G.J. Galletta, N. Vorsa, G. Jaelenkovic. 1991. 'Sunrise' highbush blueberry. *HortScience* 26(3):317-318.

Eaton, G.W. and M.G. Stewart. 1969. Blueberry blossom damage caused by bumblebees. *The Canadian Entomologist* 101:149-150.

Eck, P. 1964. Magnesium status of New Jersey blueberry plantings. *Proceedings of the 32nd Annual Blueberry Open House*, Hammonton, NJ. pp. 7-9.

_____. 1983. Optimum potassium nutritional level for production of highbush blueberry. *Journal of the American Society for Horticultural Science*. 108:520-522.

_____ . 1988. *Blueberry Science*. New Brunswick, NJ: Rutgers University Press.

Eck, P. and N.F. Childers, eds. 1966. *Blueberry Science*. New Brunswick, NJ: Rutgers University Press.

Eck, P. and C.M. Mainland. 1971. Highbush blueberry fruit set in relation to flower morphology. *HortScience* 6:494-495.

El-Agamy, S.Z.A., W.B. Sherman, and P. Lyrene. 1981. Fruit set and seed number from self- and cross-pollination highbush (4x) and rabbiteye (6x) blueberries. *Journal of the American Society for Horticultural Science 106*(4):443-445.

Erb, W.A. 1987. Evaluation of *Vaccinium* interspecific hybrids in different water deficit environments and on an upland and organic soil. PhD Dissertation, Hort. Dept., University of Maryland, College Park.

Erb, W.A., A.D. Draper, and H.J. Swartz. 1991. Combining ability for canopy growth and gas exchange of interspecific blueberries under moderate water deficit. *Journal of the American Society for Horticultural Science 116*(3):569-573.

_____ 1993. Relation between moisture stress and mineral soil tolerance in blueberries. *Journal of the American Society for Horticultural Science 118*(1):130-134.

Eynard, I. and E. Czesnik. 1989. Incidence of mycorrhiza in 4 highbush blueberry cultivars in different soils. *Acta Horticulturae* 241:115-119.

Fear, C.D., F.I. Lauer, J.J. Luby, R.L. Stucker, and C. Stushnoff. 1985. Genetic components of variance for winter injury, fall growth cessation, and off-season flowering in blueberry progenies. *Journal of the American Society for Horticultural Science 110*(2):262-266.

Filmer, R.S. and P.E. Marucci. 1964. Honeybees and blueberry pollination. *Proceedings of the 32nd Annual Blueberry Open House*. New Jersey Agricultural Experiment Station. pp. 25-27.

Finn, C.E. and J.J. Luby. 1992. Inheritance of fruit quality traits in blueberry. *Journal of the American Society for Horticultural Science 117*(4): 617-621.

Finn, C.E., J.J. Luby, C.J. Rosen, and P.D. Ascher. 1991. Evaluation *in vitro* of blueberry germplasm for higher pH tolerance. *Journal of the American Society for Horticultural Science 116*(2):312-316.

Funt, R.C., M.C. Schmittgen, and J.C. Golden. 1991. Highbush blueberry cultivar evaluation. Fruit cultural practices and cultivar trials at Overlook Farm. Ohio State University Horticulture Series 612. pp. 13-16.

Galletta, G.J. 1975. Blueberries and cranberries. In *Advances in Fruit Breeding*. Eds. J. Janick and J.N. Moore. West Lafayette, IN: Purdue University Press. pp. 154-196.

Galletta, G.J. and D.G. Himelrick. 1990. Small Fruit Crop Management. Englewood Cliffs, NJ: Prentice-Hall, Inc.

Galletta, G.J. and C.M. Mainland. 1971. Comparative effects of mechanical harvesting on highbush blueberry varieties. *Proceedings of the 1971 Highbush Blueberry Mechanization Symposium.* Eds. Mainland, C.M. and R.P. Rohrbach. North Carolina State University, Raleigh, NC. pp. 80-85.

Geohring, L. 1989. Water management. In: *Bramble Production Guide* (NRAES-35). Eds. M. Pritts and D. Handley. Ithaca, NY: Northeast Regional Agricultural Engineering Service. 93-108.

Gerber, J.F. 1970. Crop protection by heating, wind machines, and overhead irrigation. *HortScience 5*(5):428-431.

Gough, R.E. 1980. Root distribution of 'Coville' and 'Lateblue' highbush blueberry under sawdust mulch. *Journal of the American Society for Horticultural Science 105*(4):576-578.

_____ . 1982. Highbush blueberries in New England: What now—what next? *Proceedings of the New England Small Fruit School,* Concord, NH, p. 13-16.

_____ . 1983a. The occurrence of mesocarpic stone cells in the fruit of cultivated highbush blueberry. *Journal of the American Society for Horticultural Science 108*(6):1064-1067.

_____ . 1983b. Deep planting causes blueberry death. *American Fruit Grower*. October.

_____ . 1983c. Time of pruning and bloom date in cultivated highbush blueberry. *HortScience 18*(6):934-935.

_____ . 1984a. Split-root fertilizer application to potted, greenhouse-grown highbush blueberry plants. *Proceedings of the Fifth North American Blueberry Research Workers Conference*, Eds.

Crocker, T.E., and P.M. Lyrene. University of Florida, Gainesville, pp. 206-211.

_____ . 1984b. Split-root fertilizer application to highbush blueberry plants. *HortScience 19*(3):415-416.

_____ . 1992. Diagnosing disorders of the highbush blueberry. *Journal of Small Fruit and Viticulture 1*(1):63-84.

Gough, R.E. and V.G. Shutak. 1978. Anatomy and morphology of cultivated highbush blueberry. Rhode Island Agricultural Experimentation Bulletin 423.

Gough, R.E. and W. Litke. 1980. An anatomical and morphological study of abscission in highbush blueberry fruit. *Journal of the American Society for Horticultural Science 105*(3):335-341.

Gough R.E. and A. Pennucci. 1981. A trellis that's tough to top. *American Fruit Grower*, May. pp. 20 and 36.

Gough, R.E., R.J. Hindle, and V.G. Shutak. 1976. Identification of ten highbush blueberry cultivars using morphological characteristics. *HortScience 11*(5):512-514.

Gough, R.E., V.G. Shutak, and R.L. Hauke. 1978a. Growth and development of highbush blueberry. I. Vegetative growth. *Journal of the American Society for Horticultural Science 103*(1):94-97.

_____ . 1978b. Growth and development of highbush blueberry. II. Reproductive growth, histological studies. *Journal of the American Society for Horticultural Science 103*(4):476-479.

Gough, R.E., V.G. Shutak, and D.B. Wallace. 1983. Highbush blueberry culture. Rhode Island Cooperative Extension Bulletin 143.

Grout, J.M., P.E. Read, and D.K. Wildung. 1986. Influence of tissue culture and leaf-bud propagation on the growth habit of 'Northblue' blueberry. *Journal of the American Society for Horticultural Science* 111:372-375.

Gupton, C.L. 1985. Establishment of native *Vaccinium* species on a mineral soil. *HortScience* 20:673-674.

Haghighi, K. and J.F. Hancock. 1992. DNA restriction fragment length variability in the genomes of highbush blueberry. *HortScience 27*(1):44-47.

Hall, I.V. and R.A. Ludwig. 1961. The effects of photoperiod, temperature, and light intensity on the growth of the lowbush blueberry. *Canadian Journal of Botany* 39:1733-1739.

Hall, I.V., D.L. Craig, and L.E. Aalders. 1963. The effect of photoperiod on the growth and flowering of the highbush blueberry (*Vaccinium corymbosum* L.). *Proceedings of the American Society for Horticultural Science* 82:260-263.

Hancock, J.F. 1989. Blueberry research in North America. *Acta Horticulturae* 241:19-30.

Hancock, J.F. and A.D. Draper. 1989. Blueberry culture in North America. *HortScience* 24(4):551-556.

Hancock, J.F. and J.H. Siefker. 1982. Levels of inbreeding in highbush blueberry cultivars. *HortScience* 17(3):363-366.

Hancock, J.F., M. Sakin, and P.A. Callow. 1991. Heritability of flowering and harvest dates in *Vaccinium corymbosum* L. *Fruit Varieties Journal* 45(3):173-176.

Hancock, J., I. Widders, J. Nelson, and C. Schumann. 1984. Influence of foliar-applied nitrogen on blueberry yield and quality. *Proceedings of the Fifth North American Blueberry Research Workers Conference*, Gainesville, Florida.

Hancock, J.F., K. Haghighi, S.L. Krebs, J.A. Flore, and A.D. Draper. 1992. Photosynthetic heat stability in highbush blueberries and the possibility of genetic improvement. *HortScience* 27(10):1111-1112.

Hancock, J.F., J.W. Nelson, H.C. Bittenbender, P.W. Callow, J.S. Cameron, S.L. Krebs, M.P. Pritts, and C.M. Schumann. 1987. Variation among highbush blueberry cultivars in susceptibility to spring frost. *Journal of the American Society for Horticultural Science* 112(4):707-711.

Hansel, M.R., H.E. Carpenter, and C.M. Mainland. 1971. Grower evaluation of mechanization in Michigan, North Carolina, and New Jersey. *Proceedings of the 1971 Highbush Blueberry Mechanization Symposium*. (Eds.) Mainland, C.M. and R.P. Rohrbach. North Carolina State University, Raleigh, NC. pp. 24-33.

Hanson, E.J. and J.B. Retamales, 1992. Effect of nitrogen source and timing on highbush blueberry performance. *HortScience* 27(12):1265-1267.

Hapitan, J.C., V.G. Shutak, and J.T. Kitchen. 1969. Vegetative and reproductive responses of highbush blueberry to succinic acid, 2,2-dimethylhydrazide (Alar). *Journal of the American Society for Horticultural Science* 94:26-28.

Harrison, R.E., J.J. Luby, and P.D. Ascher. 1990. Effect of varying cross: self pollen ratios on the reproductive fertility of four Minnesota half-high blueberry cultivars. *Proceeding of the Sixth North American Blueberry Research and Extension Workers Conference*, Portland, Oregon.

Hartmann, H.T., A.M. Kofranek, V.E. Rubatsky, and W.J. Flocker. 1988. *Plant Science*. 2nd ed. Englewood Cliffs, NJ: Prentice-Hall.

Hedrick, U.P., ed. 1919. *Sturtevant's Notes on Edible Plants*. Albany, NY: J.B. Lyon Co.

Herath, H.M.E. and G.W. Eaton. 1968. Some effects of water table, pH, and nitrogen fertilization upon growth and composition of highbush blueberry. *Proceedings of the American Society for Horticultural Science* 92:274-283.

Hindle, R., V. Shutak, and E.P. Christopher. 1957. Relationship of wood thickness to blossoming, rate of ripening, and size of fruit on the highbush blueberry. *Proceedings of the American Society for Horticultural Science* 70:150-155.

Howell, G.S., J. Nelson, and K. Michael. 1970. Influence of pollination and nutritional status on the yield and quality of highbush blueberries. *HortScience* 5(4):326.

Howell, G.S., M.W. Kilby, and J.W. Nelson. 1972. Influence of timing of hive introduction on production of highbush blueberries. *HortScience* 7:129-131.

Howell, G.S., C.M. Hanson, H.C. Bittenbender, and S.S. Stackhouse. 1975. Rejuvenating highbush blueberries. *Journal of the American Society for Horticultural Science* 100:455-457.

Howell, G.S., B.G. Stergios, S.S. Stackhouse, and H.C. Bittenbender. 1975. Mechanical harvester vibration rates related to winter kill of highbush blueberry branches. *HortScience* 10:85-86.

Hull, J. 1967. Weed control, nutrition, and mechanical harvesting practices in the Michigan blueberry industry. International Society for Horticultural Science Working Group. Blueberry Culture. I. Symposium.

Hursalmi, H.M. 1989. Research into *Vaccinium* cultivation in Finland. *Acta Horticulturae* 241:175-184.

Janick, J., R.W. Schery, F.W. Woods, and V.W. Ruttan, eds. 1981. *Plant Science*. San Francisco, CA: W.H. Freeman and Co.

Johnston, S. 1935. Propagating low and highbush blueberry plants by means of small side shoots. *Proceedings of the American Society for Horticultural Science* 33:372-375.

_____ . 1937. Influence of cultivation on the growth and yield of blueberry plants. Michigan Agricultural Experiment Station Quarterly Bulletin 19:232-234.

_____ . 1942. Observations on the inheritance of horticulturally important characteristics in the highbush blueberry. *Proceedings of the American Society for Horticultural Science* 40:352-356.

_____ . 1943. The influence of manure on the yield and size of fruit of the highbush blueberry. Michigan Agricultural Experiment Station Quarterly Bulletin 25:374-376.

_____ . 1951. Problems associated with cultivated blueberry production in northern Michigan. Michigan Agricultural Experiment Station Bulletin 33:293-298.

Kender, W.J. and W.T. Brightwell. 1966. Environmental relationships. In *Blueberry Culture*. Eds. P. Eck and N.F. Childers. New Brunswick, NJ: Rutgers University Press. 75-93.

Knight, R.J. Jr. and D.H. Scott. 1964. Effects of temperatures on self- and cross-pollination and fruit of four highbush blueberry varieties. *Proceedings of the American Society for Horticultural Science* 85:302-306.

Korcak, R.F. 1988. Nutrition of blueberry and other calcifuges. *Horticultural Reviews* 10:183-227.

_____ . 1989a. Variation in nutrient requirements of blueberries and other calcifuges. *HortScience* 24(4):573-578.

_____ . 1989b. Aluminum relationships of highbush blueberries (*Vaccinium corymbosum* L.). *Acta Horticulturae* 214:162-166.

_____ . 1992a. Blueberry species and cultivar response to soil types. *Journal of Small Fruit and Viticulture* 1(1):11-24.

_____ . 1992b. Short-term response of blueberry to elevated soil calcium. *Journal of Small Fruit and Viticulture* 1(2):9-21.

Kramer, A., E.L. Evinger, and A.L. Shrader. 1941. Effect of mulches and fertilizers on yield and survival of the dryland and highbush blueberries. *Proceedings of the American Society for Horticultural Science* 38:455-461.

Lang, G.A. and R.G. Danka. 1991. Honey-bee-mediated cross-versus self-pollination of 'Sharpblue' blueberry increases fruit size and hastens ripening. *Journal of the American Society for Horticultural Science 116*(5):770-773.

Lang, G.A. amd E.J. Parrie. 1992. Pollen viability and vigor in hybrid southern highbush blueberries *(Vaccinium corymbosum)* L. × spp.). *HortScience 27*(5):425-427.

Lang, G.A., R.G. Danka, and E.J. Parrie. 1990. Pollen-stigma interactions and relationship to fruit development in low-chill southern highbush blueberry. *Proceedings of the Sixth North American Blueberry Research and Extension Workers Conference.* Portland, Oregon. pp. 36-44.

Lareau, M.J. 1989. Growth and productivity of highbush blueberries as affected by soil amendments, nitrogen fertilization and irrigation. *Acta Horticulturae* 241:126-131.

Leonard, R.T. 1984. Membrane associated ATPases and nutrient absorption by roots. *Advances in Plant Nutrition* 1:209-240.

Liner, H.L. 1971. A case study of blueberry mechanization economics in North Carolina. *Proceedings of the 1971 Highbush Blueberry Mechanization Symposium.* Eds. Mainland, C.M. and R.P. Rohrbach, North Carolina State University, Raleigh, NC. pp. 57-65.

Luby, J.J. 1991. Where do half-highs fit in? *Proceedings of the New England Small Fruit and Vegetable Growers Conference,* Sturbridge, Massachusetts.

Luby, J.J., D.K. Wildung, C. Stushnoff, S.T. Munson, P.E. Read, and E.E. Hoover. 1986. 'Northblue,' 'Northsky,' and 'Northcountry' blueberries. *HortScience 21*(5):1240-1242.

Lutz, J.M. and R.E. Hardenburg. 1968. The commercial storage of fruits, vegetables, and florists and nursery stocks. United States Department of Agriculture Handbook 66.

Lyrene, P.M. and W.B. Sherman. 1992. The 'Sharpblue' southern highbush blueberry. *Fruit Varieties Journal 46*(4):194-196.

Lyrene, P.M. 1992. Early defoliation reduces flower bud counts on rabbiteye blueberry. *HortScience 27*(7):783-785.

MacFarlane, R.P. 1992. An initial assessment of blueberry pollinators in New Zealand. *New Zealand Journal of Crop and Horticulture Science* 20:91-95.

Mainland, C.M. 1971. Fruit recovery and bush damage with mechanical harvesting. *Proceedings of the 1971 Highbush Blueberry Mech-*

anization Symposium. Eds. Mainland, C.M. and R.P. Rohrbach. North Carolina State University, Raleigh, NC. pp. 24-33.

_____. 1985. Commercial blueberry production guide for North Carolina. North Carolina Cooperative Extension Bulletin AG-115.

_____. 1989a. Harvesting, sorting and packing quality blueberries. *Proceedings of the 23rd Annual Open House, Southeastern Blueberry Council.* Elizabethtown, North Carolina: North Carolina State University. pp. 47-51.

_____. 1989b. Pruning. *Proceedings of the 23rd Annual Open House,* Southeastern Blueberry Council. Elizabethtown, North Carolina: North Carolina State University. 10-16.

Mainland, C.M. and P. Eck. 1969. Fruiting response of highbush blueberry to gibberellic acid under field conditions. *Journal of the American Society for Horticultural Science* 94:21-23.

Mainland, C.M. and J.P. Lilly. 1984. The effect of black plastic mulch and fertilizer treatments on the growth of highbush and rabbiteye blueberry plants. *Proceedings of the 18th National Agricultural Plastics Congress,* eds. Lamont, W. and D. Sanders. North Carolina State University, Raleigh, pp. 194-201.

Mainland, C.M., L.J. Kushman, and W.E. Ballinger. 1975. The effect of mechanical harvesting on yield, quality of fruit, and bush damage of highbush blueberry. *Journal of the American Society for Horticultural Science* 100:129-134.

Marucci, P.E. 1966. Blueberry pollination. *American Bee Journal* 106:250-251.

McCarthy, E. and R.P. Rohrbach. 1985. Harvester ground loss reduction. *Proceedings of the 19th Annual Open House, Southeastern Blueberry Council.* North Carolina State University, Raleigh, NC. pp. 84-88.

McGregor, S.E. 1976. Insect pollination of cultivated crop plants. United States Dept. of Agriculture Agricultural Handbook. 496.

Meade, J. 1989. Weed control in blueberries. *Proceedings of the Blueberry Open House,* Mount Holly, NJ.

Meade, J. 1990. Blueberry herbicide update. *Proceedings of the Blueberry Open House,* Hammonton, NJ.

Meade, J. 1991. Weed management program in blueberry. *Proceedings of the Blueberry Open House,* Atlantic City, NJ.

Meader, E.M. and G.M. Darrow. 1947. Highbush blueberry pollination experiments. *Proceedings of the American Society for Horticultural Science* 49:196-204.

Meador, D.B., C.C. Doll, J.W. Courter, R.M. Skirvin, and A.G. Otterbacher. 1983. Highbush blueberry cultural tips. *Proceedings of the Illinois Small Fruit School.* 31-34.

Merrill, T.A. 1936. Pollination of the highbush blueberry. Michigan Agricultural Experiment Station Bulletin 151.

_____. 1944. Effects of soil treatments on growth of the highbush blueberry. *Journal of Agricultural Research* 69:9-20.

Meyer, J.R. 1989. Insect and mite control. *Proceedings of the 23rd Annual Open House Southeastern Blueberry Council.* North Carolina State University, Elizabethtown, NC.

Milholland, R.D. 1982. Blueberry twig blight caused by *Phomopsis vaccinii. Plant Disease* 66:1034-1036.

Moore, J.N. 1964. Duration of receptivity to pollination of flowers of the highbush blueberry and the cultivated strawberry. *Proceedings of the American Society for Horticultural Science* 85:295-301.

Moore, J.N. and D.P. Ink. 1964. Effect of rooting medium, shading, type of cutting, and cold storage of cuttings on the propagation of highbush blueberry varieties. *Proceedings of the American Society for Horticultural Science* 85:285-294.

Moore, J.N., H.H. Bowen, and D.H. Scott. 1962. Response of highbush blueberry varieties, selections, and hybrid progenies to powdery mildew. *Proceedings of the American Society for Horticultural Science* 81:274-280.

Norvell, D.J. and J.N. Moore. 1982. An evaluation of chilling models for estimating rest requirements of highbush blueberries (*Vaccinium corymbosum* L.). *Journal of the American Society for Horticultural Science* 107(1):54-56.

Odneal, M.B. and M.L. Kaps. 1990. Fresh and aged pine bark as soil amendments for establishment of highbush blueberry. *HortScience* 25(10):1228-1229.

Parliman, B., M. Toivio, and C. Stushnoff. 1974. Propagation procedures for new half-high highbush blueberry hybrids. *Proceedings of the Third North American Blueberry Research Workers Conference.* Michigan State University, E. Lansing. pp. 17-33.

Patel, N. and J.A. Douglas. 1989. Performance of three early-bearing New Zealand highbush blueberry cultivars (*Vaccinium corymbosum* L.). *Acta Horticulturae* 241:81-86.

Patten, K., G. Nimr, E. Neuendorff, and G. Krewer. 1990. Living mulches for rabbiteye blueberry production. *Proceedings of the Sixth North American Blueberry Research and Extension Workers Conference.* pp. 121-128.

Patten, K.D., E.W. Neuendorff, G.H. Nimr, S.C. Peters, D.L. Cawthon. 1989. Growth and yield of rabbiteye blueberry as affected by orchard floor management practices and irrigation geometry. *Journal of the American Society for Horticultural Science* 114(5):728-732.

Patten, K., E. Neuendorff, G. Nimr, R. Clark, and G. Fernandez. 1991. Cold injury of southern blueberries as a function of germplasm and season of flower development. *Hortscience* 26(1):18-20.

Pavlis, G.C. 1991a. Disease and culture. The Blueberry Bulletin. Rutgers University Cooperative Extension Service VII(2):1-2.

Pavlis, G.C. 1991b. Diseases and culture. The Blueberry Bulletin. Rutgers University Cooperative Extension Service VII(7):1.

Pavlis, G.C. 1991c. Living mulches may help blueberry farmers. The Blueberry Bulletin. Rutgers University Cooperative Extension Service VII(21):3.

Peterson, D.V., C.A. Mullins, D.A. Lietzke, and D.E. Deyton. 1987. Effects of soil-applied elemental S, aluminum sulfate, and sawdust on growth of rabbiteye blueberries. *Journal of the American Society for Horticultural Science* 112:612-616.

Peterson, L.A., E.J. Stang, and M.N. Dana. 1988. Blueberry response to NH_4-N and NO_3-N. *Journal of the American Society for Horticultural Science* 113:9-12.

Phipps, C.R. 1930. Blueberry and huckleberry insects. Maine Agricultural Experiment Station Bulletin 356:107-232.

Pierre, W.H. 1933. Determination of equivalent acidity and basicity of fertilizers. Industrial Engineering Chemical Analytical Edition 5:229-234.

Ponnamperuma, F.N. 1984. Effects of flooding on soils. In T.T. Kozlowski (ed.). *Flooding and Plant Growth.* New York, New York: Academic Press. pp. 10-42.

Powell, C.L. 1982. The effect of ericoid mycorrhizal fungus Pezizella ericae (Read) on the growth and nutrition of seedlings of blueberry (*Vaccinium corymbosum* L.). *Journal of the American Society for Horticultural Science* 107:1012-1015.

Powell, C.L. and P.M. Bates. 1981. Ericoid mycorrhizas stimulate fruit yield of blueberry. *HortScience* 16:655-656.

Prange, R.K. and P.D. Lidster, 1992. Controlled atmosphere effects on blueberry maggot and lowbush blueberry fruit. *HortScience* 27(10):1094-1096.

Pritts, M. and D. Handley, eds. 1989. Bramble production guide. Northeast Regional Agricultural Engineering Service Bulletin 35.

Prokopy, R.J. and W.M. Coli. 1978. Selective traps for monitoring Rhagoletis mendax flies. *Protection Ecology* 1:45-53.

Quamme, H.A., C. Stushnoff, and C.J. Weiser. 1972. Winter hardiness of several species and cultivars in Minnesota. *HortScience* 7:500-502.

Rader, L.F., L.M. White, and C.W. Whittaker. 1943. The salt index–a measure of the effect of fertilizers on the concentration of the soil solution. *Soil Science* 55:201-218.

Read, P.E., D.K. Wildung, C.A. Hartley, and J.M. Sandahl. 1989. Field performance of in vitro-propagated 'Northblue' blueberries. *Acta Horticulturae* 241:191-194.

Retamales, J.B. and E.J. Hanson. 1989. Fate of ^{15}N-labeled urea applied to mature highbush blueberries. *Journal of the American Society for Horticultural Science* 114(6):920-923.

Rohrbach, R.P. and C.M. Mainland. 1989. Crown restrictions in blueberries reduce harvesting ground losses. *Acta Horticulturae* 241:366-372.

Rohrbach, R.P., R. Ferrell, E.D. Beasley, and J.R. Fowler. 1984. Precooling blueberries and muscadine grapes with liquid carbon dioxide. *Transactions of the American Society for Agricultural Engineering* 27:1950-1955.

Rosen, C.J., D.L. Allan, and J.J. Luby. 1990. Nitrogen form and solution pH influence growth and nutrition of two *Vaccinium* clones. *Journal of the American Society for Horticultural Science* 115(1):83-89.

Rowland, L.J. and E.L. Ogden, 1992. Use of cytokinin conjugate for efficient shoot regeneration from leaf sections of highbush blueberry. *HortScience* 27(10):1127-1129.

Russell, H.S. 1980. *Indian New England Before the Mayflower.* Hanover, NH: The University Press of New England.

_____. 1982. *A Long, Deep Furrow.* Hanover, NH: The University Press of New England.

Russell, R.S. 1977. *Plant Root Systems: Their Function and Interaction with the Soil.* London, England: McGraw-Hill.

Ryall, A.L. and W.T. Pentzer. 1974. *Handling, Transportation, and Storage of Fruits and Vegetables.* Vol. 2. Fruits and Tree Nuts. AVI Publishing Co., Inc., Westport, CT.

Safley, C.D. 1985. Harvesting, sorting, and packaging expenses. *Proceedings of the 19th Annual Open House, Southeastern Blueberry Council.* North Carolina State University, Raleigh, NC. p. 89.

Sanderson, K.R. and J.A. Cutcliffe. 1991. Effect of sawdust mulch on yields of select clones of lowbush blueberry. *Canadian Journal of Plant Science* 71:1263-1266.

Schwartze, C.D. and A.S. Myhre. 1954. Blueberry propagation. Washington Agricultural Experiment Station Circular 124.

Shelton, L.L. and J.N. Moore. 1981a. Highbush blueberry propagation under southern U.S. climatic conditions. *HortScience* 16:320-321.

_____. 1981b. Field chilling versus cold storage of highbush blueberry cuttings. *HortScience* 16:316-317.

Shelton, L.L. and B. Freeman. 1989. Blueberry cultural practices in Australia. *Acta Horticulturae* 241:250-253.

Sherman, W. and P. Lyrene. 1991. Deciduous cultivar development in Florida. *HortScience* 26(1):2,9.

Shoemaker, J.S. 1978. *Small Fruit Culture.* Westport, CT: AVI Publishing Co.

Shutak, V.G. 1966. Effect of light intensity on vegetative and reproductive growth of highbush blueberry. *Proceedings of the North American Blueberry Research Workers Conference.* Orono, Maine: University of Maine. 98-101.

_____. 1968. Effect of succinic acid 2,2-dimethylhydrazide on flower bud formation of the 'Coville' highbush blueberry. *HortScience* 3:225.

Shutak, V.G. and E.P. Christopher. 1952. Sawdust mulch for blueberries. Rhode Island Agricultural Experiment Station Bulletin 312.

Shutak, V.G., R. Hindle, and E.P. Christopher. 1956. Factors associated with ripening of highbush blueberry fruits. *Proceedings of the American Society for Horticultural Science* 68:178-183.

_____. 1957. Growth studies of the highbush blueberry. Rhode Island Agricultural Experiment Station Bulletin 339.

Shutak, V.G., R.E. Gough, and N.D. Windus. 1980. The cultivated highbush blueberry: twenty years of research. Rhode Island Agricultural Experiment Station Bulletin 428.

Siefker, J.A. and J.F. Hancock. 1987. Pruning effects on productivity and vegetative growth in the highbush blueberry. *HortScience* 22(2):210-211.

Slate, G.L. and R.C. Collison. 1942. The blueberry in New York. New York Agricultural Experiment Station Circular 189.

Smolarz, K. and A. Kostusiak. 1984. The first years of fruiting of 14 highbush blueberry cultivars in Skierniewice, Poland. *Proceedings of the Fifth North American Blueberry Research Workers Conference*. Gainesville, Florida: University of Florida. pp. 77-78.

Song, V., H.K. Kim and K.L. Yam. 1992. Respiration rate of blueberry in modified atmosphere at various temperatures. *Journal of the American Society for Horticultural Science* 117(6):925-929.

Spiers, J.M. 1980. Influence of peatmoss and irrigation on establishment of 'Tifblue' blueberry. Mississippi Agriculture and Forestry Experiment Station Research Report 4(18).

_____. 1983. Irrigation and peatmoss for the establishment of rabbiteye blueberries. *HortScience* 18(6):936-937.

Spiers, J.M. and J.H. Braswell. 1992. Soil-applied sulfur affects elemental leaf content and growth of 'Tifblue' rabbiteye blueberry. *Journal of the American Society for Horticultural Science* 117(2):230-233.

Stang, E.J., M.N. Dana, G.G. Weis, and B.H. McCown. 1990. 'Friendship' blueberry. *HortScience* 25(12):1667-1668.

Stojanov, D. 1989. Some results in bilberry, lingonberry, and highbush blueberry research in Bulgaria. *Acta Horticulturae* 241:95-99.

Strik, B.C. 1990. Bird damage control in blueberries in Oregon. *Proceedings of the Sixth North American Blueberry Research and Extension Workers Conference.* Portland, Oregon. pp. 129-132.

Takamizo, T. and N. Sugiyama. 1991. Growth responses to N forms in rabbiteye and highbush blueberries. *Journal of the Japanese Society for Horticultural Science* 60(1):41-45.

Tamada, T. 1989. Blueberry growing in Chiba-Ken, Japan. *Acta Horticulturae* 241:64-70.

Teramura, A.H., F.S. Davies, and D.W. Buchanan. 1979. Comparative photosynthesis and transpiration in excised shoots of rabbiteye blueberry. *HortScience* 14:723-724.

Townsend, L.R. 1967. Effect of ammonium N and nitrate N separately and in combination on growth of the highbush blueberry. *Canadian Journal of Plant Science* 47:555-562.

Townsend, L.R. 1973. Effects of N., P., K., and Mg on the growth and productivity of highbush blueberry. *Canadian Journal of Plant Science.* 53:161-168.

Vander Kloet, S.P. 1988. The Genus *Vaccinium* in North America. Research Branch, Agriculture Canada Publication 1828.

_____ . 1991. The consequences of mixed pollination onseed set in *Vaccinium corymbosum. Canadian Journal of Botany* 69(11): 2448-2454.

Vorsa, N. 1991. The 'Redberry' condition in the blueberry cv. Bluecrop. The Blueberry Bulletin. Rutgers University Cooperative Extension Service VII(22):3-4.

Waxman, S. 1965. Propagation of blueberries under fluorescent light at various intensities. *Proceedings of the International Plant Propagators Society* 15:154-158.

Westwood, M.N. 1988. *Temperate-Zone Pomology.* Portland, Oregon: Timber Press.

Wildung, D.K. and K. Sargent. 1989a. The effect of snow depth on winter survival and productivity of Minnesota blueberries. *Acta Horticulturae* 241:232-237.

_____ . 1989b. Effect of rowcovers on the winter survival and productivity of Minnesota blueberries. *Acta Horticulturae* 241: 238-243.

Wildung, D.K., K. Sargent, and C. Rosen. 1990. Soil acidification of Minnesota soil for blueberry production. *Proceedings of the*

Sixth North American Blueberry Research and Extension Workers Conference, Portland, Oregon.

Williams, R. 1643. *A Key Into the Language of America*. London, England: G. Dexter.

Windus, N.D., V.G. Shutak, and R.E. Gough. 1976. CO_2 and C_2H_4 evolution by highbush blueberry fruit. *HortScience* 11:515-517.

Wright, G.C., K.D. Patten, and M.C. Drew. 1992. Salinity and supplemental calcium influence growth of rabbiteye and southern highbush blueberry. *Journal of the American Society for Horticultural Science* 177(5):749-756.

Appendix 1

Conversion Table

Col. 1 × this = Col 2	Col 1	Col 2	Col. 2 × this = Col 1
		Length	
2.54	inches	centimeters	0.3937
0.3048	feet	meters	3.2810
1.609	miles	kilometers	0.6214
30.48	feet	centimeters	0.3280
0.9144	yards	meters	1.0940
		Area	
0.4047	acres	hectares	2.4710
6.4520	sq in	sq cm	0.1050
0.8	sq yd	sq m	1.2000
0.4	sq mi	sq km	2.6000
0.004	acres	sq km	247.0000
0.093	sq ft	sq m	10.8000
		Volume	
102.80	acre–in	cu in	0.0097
0.9463	qt (liq US)	liters	1.0570

Volume (continued)

1.136	qt (Imp)	liters	0.8799
3.785	gal (US)	liters	0.2642
4.546	gal (Imp)	liters	0.2200
29.57	oz (liq US)	milliliters	0.0338
4.9	teas	milliliters	0.2000
15	tbl	milliliters	0.0670
0.24	cups	liters	4.1700
0.47	pints	liters	2.1300
0.03	cu ft	cu m	35.0000
0.2832	cu ft	hectoliters	3.5320
0.76	cu yd	cu m	1.3000
0.352	bushel	hectoliters	2.8400

Weight

28.35	oz (avoir)	grams	0.0353
0.4536	lb (avoir)	kilograms	2.2050
0.4540	cwt	quintal	2.2050
907.2	tons (short)	kilograms	0.0011
0.9072	tons (short)	tonne	1.1020

Pressure and Head

6.90	lb/sq in	kilopascals	0.1450
0.100	bars	megapascals	10.0000
0.069	lb/sq in	bars	14.5000
1.013	atmospheres	bars	0.9870

Pressure and Head (continued)

1.033	atmospheres	kg/sq cm	0.9680
0.0703	lb/sq in	kg/sq cm	14.2200

Rates

1.12	lb/acre	kg/ha	0.8920
9.35	gal/acre	L/ha	0.1070
0.120	lb/gal	kg/L	8.3300
7.49	oz/gal	g/L	0.1340
2.24	tons/acre	tonnes/ha	0.4460
0.89	bu/acre	hl/ha	1.1500
1.12	cwt/acre	q/ha	0.8920
67.2	bu/acre	kg/ha	0.1500

Water Measurement

0.1233	acre–ft	ha–m	8.1090
0.01028	acre–in	ha–m	97.2900
12.33	acre–ft	ha–cm	0.0811
1.028	acre–in	ha–cm	0.9730
102.8	acre–in	cu m	0.0097
1.0194	cu ft/s	ha–cm/hr	0.9810
0.00227	gal/min	ha–cm/hr	440.3000
101.94	cu ft/s	cu m/hr	0.0098
0.227	gal/min	cu m/hr	4.4030

Light

10.764	ft–c	lux	0.0929

Equivalents

Dry Measure

16 oz = 1 lb

1 ton = 2000 lb

1 pt = 33.6 cu in

1 gal = 268.8 cu in

1 peck = 537.6 cu in

1 peck = 16 pt

1 peck = 8 qt

1 bu = 4 pecks

1 bu = 64 pts

Liquid Measure

1 tbls = 3 teas

1 oz = 2 tbls

1 c = 8 oz

1 pt = 2c

1 pt = 29 cu in

1 pt = 16 oz

1 qt = 2 pt

1 qt = 32 oz

1 gal = 4 qt

Square Measure

1 acre = 43560 sq ft

1 acre = 208.7 ft sq

1 sq yd = 9 sq ft

1 sq mi = 1 section

1 acre = 4840 sq yd

1 sq ft = 144 sq in

1 sq mi = 640 acres

Concentration

1 ppm = % $\times$ 10000

1 ppm = 1 mg/L

1 ppm = 1 lb/100000 gal water

1 ppm = 1/1000000

1 ppm = 1 mg/kg

1 ppm = 17 $\times$ grains/gal

1 ppm = 0.64 $\times$ micromohs/cm (100 − 5000)

1 ppm = 640 $\times$ millimohs/cm (0.1 − 5.0)

1 millimoh = 1000 micromohs

millimohs/cm = EC $\times$ 11000 @ 25C

1 millimoh = 10 meq/L

Concentration (continued)

micromohs/cm = EC $\times$ 101000000 @ 25C

1000 micromohs/cm = 700 ppm

1000 micromohs/cm = 1 teas salt/a–ft water

Linear Measure

1 ft = 12 in	1 yd = 3 ft
1 rd = 16.5 ft	1 rd = 198 in
1 rd = 5.5 yd	1 mi = 5280 ft
1 mi = 320 rd	1 mi = 1760 yd
1 rd $\times$ 1 mi = 2 acres	

Cubic Measure

1 cu ft = 1728 cu in	1 cu ft = 0.037 cu yd
1 cu ft = 7.4805 gal	1 cu yd = 27 cu ft
1 cu yd = 46656 cu in	1 cu yd = 202 gal
1 gal = 128 oz	1 gal = 8 pt

Pressure and Head

1 atm = 14.7 psi	1 amt = 29.9 in Hg
1 atm = 33.9 ft water	1 ft water = 0.8826 in Hg
1 ft water = 0.4335 psi	1 in Hg = 1.133 ft water
1 in Hg = 0.4912 psi	1 in water = 0.0736 in Hg
1 in water = 0.03613 psi	1 psi = 2.307 ft water
1 psi = 2.036 in Hg	

Flow (US)

1 cu ft water/s = 1 s–ft

1 a–in/hr = 448.8 gal

1 s–ft = 7.5 gal/s

1 s–ft for 1 hr = 1 a–in

1 a–in/hr = 1 s–ft

1 s–ft = 3600 cu ft/hr

1 s–ft for 12 hr = 1 a–ft

1 s–ft for 1 day = 1.98 a–ft

Power

1 horsepower = 550 ft–lb/s   1 horsepower = 33000 ft–lb/min

1 horsepower = 0.7457 kwt   1 kwt = 1.314 horsepower

1 horsepower–hr = 0.7457 kwt–hr

1 a–ft water lifted 1 ft = 1.372 horsepower–hr work

1 a–ft water lifted 1 ft = 1.025 kwt–hr work

Miscellaneous Equivalents

1 oz/sq ft = 2720.7 lb/a

1 oz/sq yd = 302.3 lb/a

1 oz/100 sq ft = 27.2 lb/a

1 lb/100 sq ft = 436.6 lb/a

1 lb/1000 sq ft = 43.6 lb/a

1 gal/a = 5.8 tbls/1000 sq ft

50 gal/a = 9.2 pt/1000 sq ft

100 gal/a = 2.3 gal/1000 sq ft

100 lb/a = 2.3 lb/1000 sq ft

1 cu m/s = 35.14 cu ft/s

1 cu m/hr = 0.278 L/s

1 L/s = 15.852 gal/min (US)

Miscellaneous Equivalents (continued)

1 L/s = 3.6 cu m/hr

1 cu ft/s = 28.32 L/s (US)

1 cu ft/s = 448.8 gal/min (US)

1 cu ft/s = 1 a–in/hr (approx)

1 cu ft/s = 2 a–ft/day (approx)

1 US gal/min = 0.06309 L/s

1 a–in = 3630 cu ft

1 a–in = 102.8 cu m

1 a–in water = 113 t

1 a–in water = 27154 gal

1 a–ft soil = 2000 t

1 a–ft = 1233.5 cu m

1 a–ft = 43560 cu ft

1 a–ft water = 325851 gal

1 a–ft water = 2722500 lb

1 cu ft water = 62.4 lb

1 cu ft water = 7.48 gal

1 cu ft muck = 25–30 lb

1 cu ft clay or silt = 68–80 lb

1 cu ft sand = 100–110 lb

1 cu ft loam = 80–95 lb

1 cu ft average soil = 80–90 lb

Soil surface plow depth (6.6 in) = 1000 t/a

The volume of compact soil increases about 20% when tilled.

Miscellaneous Equivalents (continued)

1 mph = 88 fpm

1 mph = 1.467 fps

2 mph = 174 fpm

3 mph = 264 fpm

4 mph = 352 fpm

1 fpm = 0.01667 fps

1 fpm = 0.01136 mph

Abbreviations

ounce–oz	ton–t	inch–in
yard–yd	meter–m	hectare–ha
foot–ft	rod–rd	mile–mi
teaspoon–teas	cup–c	tablespoon–tbl
gallon–gal	pint–pt	quart–qt
acre–a	micrometer–um	millimeter–mm
gram–gm	kilometer–km	centimeter–cm
milligram–mg	kilogram–kg	centimeter–cm
milliliter–ml	hectoliter–hl	liter–L
kilopascals–kPa	megapascals–MPa	square–sq
pound–lb	miles/hour–mph	feet/minute–fpm
cubic–cu	quintal–q	feet/second–fps
peck–pk	bushel–bu	hundredweight–cwt
second–s	minute–min	second–foot=s–ft
foot–pound=ft–lb	kilowatt–kwt	milliequivalent–meq
lb/sq in–psi		electrical conductivity–EC

Appendix 2

Miscellaneous Information Useful to the Grower

Approximate Equal Amounts of Material
for Various Amounts of Water

Water (gal)	Amount of Material (Dry Measure)					
100	1 lb	2 lb	3 lb	4 lb	5 lb	6 lb
50	8 oz	1 lb	1.5 lb	2 lb	2.5 lb	3 lb
5	3 tbsp	1.5 oz	2.5 oz	3.3 oz	4 oz	5 oz
1	1 tsp	1 tbsp	1.5 tbsp	2 tbsp	3 tbsp	3 tbsp

Fluid Measure

100	0.5 pt	1 pt	2 pt	3 pt	4 pt	5 pt
50	4 oz	8 oz	1 pt	24 oz	1 qt	2.5 pt
5	1 tbsp	1 oz	1.5 oz	2.5 oz	3 oz	4 oz
1	5 tsp	1 tsp	2 tsp	3 tsp	4 tsp	5 tsp

Length of Row Required for One Acre

Row Spacing		Length	
(in)	(cm)	(vd)	(m)
36	90	4840	4404
48	120	3630	3303
60	150	2904	2643

72	180	2420	2202
84	210	2074	1887
96	240	1815	1652
108	270	1613	1468
120	300	1452	1321
132	330	1320	1201
144	360	1210	1101

Number of Plants Required Per Acre at Various Spacings in Feet (Does Not Account for Headland)

Spacing (ft)	Plants	Spacing (ft)	Plants
3×6	2420	8× 9	605
4×6	1815	9× 9	537
6×6	1210	3×10	1452
4×8	1361	4×10	1089
6×8	908	5×10	871
8×8	680	6×10	726
4×9	1210	8×10	544
6×9	806	10×10	435

How to Determine Ground Speed Without an Indicator

Set 2 markers 88 feet apart, then select gear and throttle setting with which you'll be spraying, take a running start and note time needed to travel the distance. Since 88 feet is equal to 1/60 of a mile, divide the time of travel into 60 to calculate speed in miles per hour. For example, if it takes 12 seconds to travel 88 feet, then field speed is 60/12 = 5 mph.

How to Calibrate a Sprayer

Calibrate only when sprayer is running smoothly. If a calibration jar is available, attach it beneath a nozzle, travel the predetermined test run, and read the rate of spray per acre from the jar scale. Without a collection jar, measure the amount of spray emitted from the nozzle for one minute. Determine travel speed in feet per minute. Use the following formula to determine amount applied per acre:

Nozzle flow $\times$ 43560/ speed $\times$ pattern width = gal/acre.

Example: 0.23 gal/min recorded flow rate
               310 feet/min recorded ground speed
               18-in nozzle width

0.23 $\times$ 43560/310 $\times$ 18/12 = 21.54 gal/min

How to Calibrate Granular Applicators

Use a wide-mouthed jar calibrated in fluid ounces. Attach jar to a delivery tube and drive 1000 feet at normal operational field speed. Even off granules and read amount. This is what the spreader is delivering at its present setting. Find the rate per acre to be applied in the following table and multiply the bandwidth (in) by the factor opposite that number.

Broadcast Rate (lb/acre)	Factor
10	0.426
15	0.639
20	0.852
25	1.065
30	1.278
35	1.491
40	1.704
45	1.917
50	2.130

Example: To apply 40 lb/acre in a 16-in band, each spout must deliver 27.264 oz every 1000 feet.

(16 in width $\times$ 1.704 = 27.264 oz)

Adjust the spreader to deliver the calculated amount of material.

Appendix 3

World Production
of the Highbush Blueberry

UNITED STATES

The highbush blueberry is a North American fruit, and the United States leads the world in production (Table A3-1). Michigan, New Jersey, the Pacific Northwest, and North Carolina produce most of the crop. Total production now exceeds 150 million (67.5 million kg) each year. Nearly all of this is sold on the fresh market.

AUSTRALIA AND NEW ZEALAND

Australia has about 700 acres (280 ha) of blueberries in production (Clayton-Green 1989). This includes rabbiteye, highbush, and southern highbush (mainly 'Sharpblue'). Plantings range in age from current-year to ten years old. There were a great number of new plantings during the 1980s, particularly by small, part-time growers, but recently the number has decreased substantially due to high interest rates and some market protectionism. Blueberry production is year-round, with the peak season late in the year. The low-chill types planted on the north coast of New South Wales are harvested in October and November; early highbush in Northern Victoria and rabbiteye cultivars in New South Wales, in December; and standard highbush in Victoria, in January. New rabbiteye plantings in Victoria have extended the season into February. About 880-1100 tons (800-1000 tonnes) were produced during the 1990-91 season, with one company producing about 60% of the total crop. Although 95% of the production comes from coastal areas of only two states, Victoria and New South Wales, small

Table Appendix 3-1. Highbush blueberry production in North America.

Region	1989 (000 Pounds)	1985–1989 Mean
Michigan	60,100	53,357
New Jersey	40,000	34,800
British Columbia	16,000	14,370
North Carolina	10,000	8,440
Washington	6,300	5,520
Oregon	9,400	6,600
Georgia	2,600	3,260
Arkansas[z]	2,900	2,480
Ontario	1,960	1,571
Cultivated Total	149,260	130,083

[z]Includes Mississippi, Louisiana, and Florida
Source: North American Blueberry Council

quantities are also produced in Tasmania, South Australia, and Western Australia.

The most popular cultivars in production are the 'Sharpblue' and 'Bluecrop,' with some 'Spartan' and 'Earliblue.' The Knoxfield selections 'Brigitta,' 'Rose,' and 'Denise' are also popular. Growers are encouraged to extend the harvest season by planting earlier cultivars on early sites and later cultivars on later sites. Most production occurs on upland soils. Incidence of *Phytophthora*, a problem in early plantings, has been reduced through improved management and planting on raised beds (Shelton and Freeman 1989). Birds continue to be a major problem for growers. Parrots are particularly troublesome since, in addition to destroying the fruit, they also eat the flowers, green fruit, and even the netting. The high cost and low availability of labor; the lack of consumer familiarity with the product; the easy availability of other, less expensive fruit (such as lychees, mangoes, and pineapples); and problems with marketing and promotion have all plagued the blueberry industry in Australia. In addition, there is some competition from other countries in the southern hemisphere, notably New Zealand, where more than 1250 acres (500 ha) are presently in production in areas of the North

Island. Popular cultivars there include 'Atlantic,' 'Jersey,' 'Dixi,' and 'Burlington,' which unfortunately ripen during the market slump of mid-December. New cultivars ripening with 'Earliblue' have been developed by New Zealand researchers. 'Puru,' 'Nui,' and 'Reka' are all seedlings of 'E 118' [('Ashworth' × 'Earliblue') × 'Bluecrop'], and they have better quality and higher yield than 'Earliblue.' Further, they all possess some ability to produce light autumn crops on current season's wood. Such crops can be profitable, since they ripen two to four weeks after the last rabbiteyes (Patel and Douglas 1989). (Clayton-Green, personal communication)

JAPAN

Expansion of the industry from experimental to commercial is underway in this country. The main cultivars presently in production include 'Weymouth,' 'Collins,' 'Blueray,' 'Bluecrop,' 'Berkeley,' 'Dixi,' and 'Herbert.' Harvest begins in early June and extends over four to seven weeks. Fruit quality has been low, due to the cloudy and rainy conditions prevailing during ripening and harvest (Tamada 1989).

EUROPE

Spain

Researchers are still experimenting with this crop in Spain. 'Patriot' and 'Blueray' appear to grow most vigorously, while 'Earliblue' performs quite poorly. (Arzuaga, personal communication)

Holland

Holland has about 60 growers with 500 acres (200 ha) in production. The harvest begins with 'Bluetta' in late June and ends with 'Elliott' in mid-September. The best-producing cultivars are 'Berkeley,' 'No. 139,' '1613-A,' 'Elliott,' 'G-71,' and 'Nelson' (Dijkstra and Wijsmuller 1989). In 1989, this small country exported 774,400 lb (352,000 kg) of berries, while its home-market consumption amounted to 484,000 lb (220,000 kg). Growers supplied 937,200 lb (426,000 kg) for the fresh market in 1990. (Wijsmuller, personal communication)

Germany

The 1235 acres (500 ha) of berries in Germany are mostly produced in plantings of 0.5-0.9 acres (1-2 ha) in size. The harvest season begins in mid-July with 'Earliblue' and ends in southern areas with 'Elliott' in mid-September. 'Bluecrop' is the best cultivar for the first half of the season, and cultivars ripening after 'Darrow' are too late for northern areas. 'Weymouth' is gradually being replaced by 'Bluetta.' The German cultivars 'Ama' and 'Heerma' are being planted for mechanical harvesting (Blasing 1989a). Plants are rarely sprayed and receive low amounts of fertilizer. Most plantations are mulched with pine sawdust and are pruned no sooner than six years after planting. Placement of three hives per hectare is usually necessary for good pollination. Fruit are hand-harvested by pickers who receive about $0.45 per lb ($1 per kg), and the product is sold mainly through one wholesaler for about $1.80 per lb ($4 per kg). Presently, the mean annual per capita consumption of this fruit is 1.2-1.5 oz (40-50 g) per year, but there is a strong marketing campaign underway to increase this. (Bunnemann, personal communication)

Finland

Presently, cultivation of this crop is successful only in southwest Finland (Hursalmi 1989). Best-producing cultivars have been 'June' and 'Rancocas,' but their susceptibility to *Fusicoccum* canker and their poor winter hardiness have interfered with production. The recently introduced Finnish cultivar 'Aron' [('Rancocas' × (*V. uliginosum* × 'Rancocas')] is very cold-hardy and canker-resistant. The bush is about 3 ft (1 m) high and produces moderate yields of good-quality, medium-sized fruit.

Norway

Commercial production in this country is just getting started. There are about 2.2 acres (1 ha) of berries presently under evaluation in plantings outside Oslo (Blasing 1989b).

Romania, Former Yugoslavia, and Bulgaria

These countries have small commercial plantings of American and European cultivars. In Bulgaria, all cultivars ripen simultaneously. 'Coville' is the most productive, followed by 'Earliblue,' 'Jersey,' 'Goldtraube,' and 'Concord' (Stojanov 1989).

Poland

This country has small-scale production that amounts to about 220 acres (100 ha) nationwide. There are also a number of home plantings. 'Earliblue' is the first cultivar to be harvested, followed by 'Weymouth,' 'Ivanhoe,' 'Herbert' and 'Jersey' (Smolarz and Kostusiak 1984).

SOUTH AMERICA

Chile

There are presently about 150 acres (60 ha) of both highbush and rabbiteye blueberries in production, with an additional 50 acres (20 ha) recently planted. About 22,000 lb (10,000 kg) of highbush berries were exported to the United States during the 1991 season. Plantings are concentrated in the southern part of the country and fruit ripen during December and January, with 'Stanley' and 'Bluecrop' producing the highest yields. Plant establishment is difficult in some areas because of inadequate soils (high pH) and, possibly, the lack of ericaceous mycorrhizae. Other problems include the fact that all cultivars bloom simultaneously and that over half are harvested within a two-week period corresponding to the post-Christmas slump in sales. The industry is notable in that it is based almost entirely upon micropropagated plants. At present, the blueberry crop has no major pests and suffers only very light bird damage. (Munoz S., personal communication)

Appendix 4

Diagnosing Disorders
of the Highbush Blueberry

Establishing and caring for the highbush blueberry plantation requires considerable effort and expense. Inevitably, all plantings develop problems that, if not corrected, can result in partial or complete loss of a crop or the entire plantation. Further, the grower often sees little direct sign of the cause of the problem. Certainly, if insects are seen chewing the leaves, they can be identified and controlled. But often, problems are environmentally induced or result from injuries to the root system, which are difficult to actually see. The grower then sees only the symptoms and not the actual sign or cause. Proper diagnosis is essential to correcting the problem.

Nutritional deficiencies strongly influence the growth of the plant, though only a few of the elements required for proper plant growth have ever been found deficient under field conditions. Likewise, hundreds of insects, pathogens, and other pests can cause slight damage over a wide area; a few can become major problems in localized areas. Of course, growers should pay particular attention to major pests in their areas. For example, Australian growers are faced with particularly troublesome parrots that not only eat the fruit and flowers but also the netting over the plants. Further, all growers should become familiar with pests generally indigenous to blueberry-producing areas throughout the world, since they can expect to encounter one or more of these in their planting.

Environmental problems, such as winter injury, hardpan, and wind damage affect growers everywhere, and everyone involved in growing blueberries should learn to recognize their symptoms. These are often general in nature and result in shoot breakage and stunting or discoloration of leaves over the entire plant.

The following diagnostic key will allow the grower to work backward in identifying common problems (Gough 1992). An analysis of the injury symptoms might then lead to identification of the cause of injury. However, because damage from a number of causes may show similar symptoms, and because symptoms of one problem might mask those of another, the grower must remember that there often is not a simple answer or remedy, and that all possible causes must be systematically evaluated and appropriately dismissed to arrive at the true cause of the damage. Each type of injury is then discussed further to aid the grower in arriving at final identification of the problem.

Complete remedies are not discussed since recommendations, particularly those for control of pests, vary considerably over time and by location. Once the probable cause has been determined, the grower should always consult local Cooperative Extension Service personnel or state specialists for verification and recommended corrective procedures.

ANALYTIC KEY TO BLUEBERRY DISORDERS

Symptoms	Possible Causes	Appendix 4 Reference Number
I. Entire plant affected		
A. Weak growth of entire top	1. N deficiency	3.1
	2. Moisture deficiency	1.1
	3. Hardpan	1.2
	4. Root damage	2.1
	5. Crown damage	2.2
	6. Rodent damage	4.1
	7. Virus	4.4.1-5
	8. Herbicide damage	2.3
	9. Borers	4.5.1
	10. Root weevils	4.5.2

Symptoms	Possible Causes	Appendix 4 Reference Number
	11. Japanese beetle	4.5.3
	12. Nematodes	4.3
	13. Powdery mildew	4.4.16
B. Weak growth of individual canes or branches	1. Crown damage	2.2
	2. Root damage	2.1
	3. Borers	4.5.1
	4. Rodent damage	4.1
	5. Herbicide damage	2.3
	6. Root weevils	4.5.2
	7. Japanese beetle	4.5.3
C. Cane dieback	1. Canker	4.4.6-9
	2. Witches' Broom	4.4.10
	3. Crown gall	4.4.11
	4. *Phytophthora* root rot	4.4.17
	5. Winter damage	1.3
	6. Herbicide damage	2.3
	7. Borers	4.5.1
	8. Root weevils	4.5.2
	9. Japanese beetle	4.5.3
II. Shoot damage		
A. Dieback of one-year-old stem in late winter to early summer	1. Winter damage	1.3
B. Holes in shoots, with frass	1. Borers	4.5.1

Symptoms	Possible Causes	Appendix 4 Reference Number
C. Shoot-tip dieback	1. K deficiency	3.3
	2. Borers	4.5.1
	3. Cold damage	1.4
	4. *Phomopsis*	4.4.18
	5. *Botrytis*	4.4.15
	6. Anthracnose	4.4.12
	7. Mummyberry	4.4.13
D. Stem dieback	1. Stem blight	4.4.14
	2. Necrotic ringspot	4.4.2
	3. Wind damage	1.6
III. Leaves		
A. Somewhat dwarfed, but not curled; off-color green	1. N deficiency	3.1
	2. Mn deficiency	3.2
	3. Moisture deficiency	1.1
	4. Hardpan	1.2
	5. Root damage	2.1
	6. Crown damage	2.2
	7. Cold damage	1.4
	8. Fe deficiency	3.5
	9. Stem blight	4.4.14
B. Somewhat curled, dwarfed, often marginally scorched; center green. May have some necrotic spotting	1. Cold damage	1.4
	2. Spray damage	2.4
	3. Fertilizer burn	2.5
	4. K deficiency	3.3
	5. Mn deficiency	3.2
	6. Mn toxicity	3.6
	7. Herbicide damage	2.3
C. Narrow, wavy, straplike; midrib red	1. Shoestring	4.4.1

Symptoms	Possible Causes	Appendix 4 Reference Number
D. Strongly distorted, dwarfed, chlorotic spots, rings, or lines	1. Necrotic ringspot	4.4.2
E. Leaves spotted, sometimes chlorotic, sometimes falling prematurely	1. Spray damage 2. Foliar fertilizer damage 3. Red ringspot 4. Anthracnose 5. Herbicide damage	2.4 2.8 4.4.3 4.4.12 2.3
F. Margins red or yellow, interveinal chlorosis	1. Mg deficiency 2. Stunt	3.4 4.4.4
G. Leaves whitened; red-bordered chlorotic spots on upper surface	1. Powdery mildew	4.4.16
H. Leaves spotted	1. Leaf spots 2. Spray damage 3. Foliar fertilizer damage	4.4.19-20 2.4 2.8
I. Leaves rolled or tunneled	1. Leaf miner	4.5.11
J. Leaves light green, becoming red/yellow mottled	1. Mosaic	4.4.5

Symptoms	Possible Causes	Appendix 4 Reference Number
IV. Flower Buds		
A. Dead before swelling	1. Winter damage	1.3
	2. Bud mite	4.5.4
B. Dead after swelling, but before opening	1. Cold damage	1.4
C. Missing after setting	1. Bird damage	4.2
V. Blossoms		
A. Blighting	1. *Botrytis*	4.4.15
	2. Anthracnose	4.4.12
B. Wine-colored; remain attached for 10-14 days	1. Poor pollination	1.5
	2. Blossom weevil	4.5.8
C. Fruit not setting properly	1. Poor pollination	1.5
	2. N deficiency	3.1
D. Small, brown holes in unopened blossom clusters	1. Blossom weevil	4.5.8
	2. Bumblebees	4.5.10
VI. Fruit		
A. Wormy	1. Blueberry maggot	4.5.5
	2. Fruitworms	4.5.6
B. Shriveled	1. Mummyberry	4.4.13
	2. *Botrytis*	4.4.15
	3. Crown damage	2.2
	4. Bird damage	4.2
	5. Anthracnose	4.4.12
	6. Cherry fruitworm	4.5.6
	7. Plum curculio	4.5.7

Symptoms	Possible Causes	Appendix 4 Reference Number
C. Undersized	1. Overbearing	2.7
	2. Improper pruning	2.6
	3. Moisture deficiency	1.1
	4. Root damage	2.1
	5. Crown damage	2.2
D. Red-specked	1. Putnam scale	4.5.9
E. Soft; leaky	1. Blueberry maggot	4.5.5

DESCRIPTIONS AND DISCUSSIONS

1.1. Moisture Deficiency

The size of leaves is often reduced over the entire plant. Crinkling and curling usually does not occur, but leaves may appear a dull, deep grayish green or, in severe cases, a bright red. Shoot growth is reduced, and the bottom of leaf petioles often appears red.

Moisture deficiency can be the result of actual drought or the plant's inability to take up water due to root or crown damage. The reduced growth incident with this is compounded by subsequent nutrient deficiencies brought about by the plant's inability to absorb and translocate nutrients present in the soil.

1.2. Hardpan

A nearly impervious layer of compacted soil existing usually less than 12 in (30 cm) below ground level will restrict root penetration and, therefore, soil volume available to the root system. This will in turn restrict the plant's uptake of water and nutrients.

Drought and/or deficiency symptoms usually appear during summer months, when the crop is ripening and the plant is stressed (see 1.1). Similar symptoms can also appear when good soil is underlaid with sand, gravel, or rock.

1.3. Winter Damage

Winter damage may be accompanied by several symptoms. Damage to the root system caused by extreme cold can result in drought and/or deficiency symptoms in the top of the plant. Heaving–in which plants are pushed up out of the ground by alternate freezing and thawing of the soil moisture–can be a particular problem on young plants that have not yet established good root systems.

Dead shoot tips and bull canes are commonly a result of winter damage. Severe late-fall or early-winter freezes often result in dieback of those shoots that continued growth into late summer and early fall and were consequently hardened insufficiently. Damage resembles that caused by *Phomopsis* (see 4.4.18).

Severely cold winter temperatures below –20°F (–29°C) can cause dieback of even hardened shoots, or the destruction of flower buds without shoot dieback.

Warm conditions in early spring can result in the uppermost few flower buds on a shoot swelling somewhat, perhaps even opening slightly, then dieing when temperatures fall.

Winter damage sometimes will not become evident until warmer weather in late spring and is therefore often not noticed during pruning operations. An obvious exception to this is snow damage, which results in cane breakage.

1.4. Cold Damage

Cool spring temperatures often will cause plants to take on a purplish or yellowish cast. These symptoms may last for a few weeks, then disappear with the onset of warmer temperatures.

Spring temperatures below freezing can cause some damage to new shoots and flowers. Temperatures around 25°F (–4°C) may cause new shoots and leaves to blacken and die. Slightly higher temperatures will result in dwarfed and curled leaves. Swollen flower buds may be killed by temperatures around 20°F (–7°C), while those in bloom will tolerate temperatures of only about 30°F (–1°C). The tip bud on a shoot is usually the least hardy, followed by the second, third, etc. Within a bud, the tip flower of the cluster is the least hardy, followed by the second, third, etc. Injured flowers and buds blacken and die.

Cold temperatures during bloom also affect production indirectly by discouraging bee flight and hampering pollination and fertilization.

1.5. Poor (Incomplete) Pollination

This can result in a substantial decrease in total fruit yield. It is mainly attributed to reduced bee activity as a result of adverse weather conditions, low native populations, or use of certain insecticides. Bee activity is reduced when: temperatures fall below 55°F (13°C), wind speeds exceed 15 mph, or rainy conditions exist. Rain and extreme temperatures can also affect pollination by destroying pollen on the stigma or preventing it from germination.

1.6. Wind Damage

Strong winds can cause severe mechanical damage to young shoots, which wilt suddenly after the storm. Look for broken stems following high winds.

2.1. Root Damage

Weak growth in the upper portion of the plant is often the first visible symptom of root damage. Roots may suffer damage in severely cold winters, when soil temperatures drop below –20°F (–29°C). This is especially true when there is a lack of snow cover, vegetation, or mulch–all of which insulate against extreme cold. Under these conditions, roots near the surface will be killed, while deeper roots remain unaffected.

Damage can also be caused by excessive or improper fertilizer application (see 2.5), a rapid rise in the water table, very deep cultivation, rodent damage (see 4.1), or extreme drought conditions (see 1.1). Fertilizer applied in narrow, concentrated bands, especially if applied close to the crown, can burn root tissue by rapid dehydration. Roots will subsequently die, and the absorptive capacity of the plant will be severely restricted. A rapid rise in the water table will result in deeper roots being killed, while those near the surface remain functional. Very high water tables, as well as shallow hardpan (see 1.2), result in a shallow root system more prone to winter and drought damage. Excessively deep cultivation will sever feeder roots and result in weak growth and some dieback, particularly under stress conditions. Improper cultivation on one side of the plant may result in symptoms on that side only.

2.2. Crown Damage

This is usually the result of cultivating too closely to the bush or of improper application of fertilizer (see 2.5) or herbicide (see 2.3). It can also be caused by borers (see 4.5.1), gall (see 4.4.11), rodents (see 4.1), and winter damage (see 1.3).

2.3. Herbicide Damage

Symptoms usually occur one time only, and do not progressively worsen. Damage becomes apparent a few days after herbicide application. It is often worse in some rows, or at row ends where the sprayer may have started, stopped, or slowed.

Simizine Damage

Leaves are chlorotic, small, and straplike, resembling leaves of a weeping willow.

Paraquat Damage

Leaves have coffee-brown necrotic spots where spray droplets contacted the surface. Spots are bordered by a chocolate-brown-to-black band and they may merge into each other to form irregular patches of dead tissue, particularly along the midrib. Leaf margins, and especially the leaf tip (where spray accumulates), are almost always involved.

Glyphosate Damage

Leaves will brown, then blacken and fall. Shoots will be generally unaffected, except for some slight tip dieback; they will appear as though leaves had been stripped off by hand.

2.4. Spray Damage

This type of damage occurs when pesticides or foliar nutrient sprays are applied at excessive concentration, when the relative humidity is very high, or when air temperatures are above 75°-80°F (24°-27°C). Some damage also occurs when dormant oil sprays are applied after leaf emergence.

Typical spray burn appears as a scorching of the leaf margins and tip, with the center often remaining green. Both new and old leaves are affected, with little or no curling.

2.5. Fertilizer Burn

When excessive amounts of commercial fertilizer are dropped in clumps, spread in a concentrated band, or applied too close to the plant, the foliage can become scorched. The symptoms are similar to those of spray burn, except that new foliage often shows greater injury. If the fertilizer has repeatedly been applied in a narrow, concentrated band, then the root system will often not spread beyond that band, making the plant more susceptible to drought damage. This could possibly result in shoot dieback, because of the decreased capacity of the roots to supply the top.

2.6. Improper Pruning

Failure either to prune adequately enough or to remove enough flower buds on cultivars that form excessive numbers can result in overly abundant fruit set and development of small fruit as a result of competition for nutrients and water.

2.7. Overbearing

This results in very small, low-quality, late-ripening fruit. It can be caused by improper pruning (see 2.6) or by improper use of certain growth regulators that induce parthenocarpy. Use of these gibberellin-type growth regulators following adequate bee activity during bloom can result in considerable overbearing.

2.8. Foliar Fertilizer Damage

This is a result of application of foliar fertilizer to the leaves in too high a concentration or in temperatures above 85°F (29°C)—especially in the presence of high relative humidity. Usually, spotting and marginal scorch appear, and damage resembles spray burn.

3.1. Nitrogen Deficiency

Plant growth appears stunted. Shoots grow only a few inches, and leaves are dwarfed and a uniform pale green. In moderately severe cases, leaves become chlorotic, and berry size and yield are reduced substantially. In severe cases, leaves redden and eventually become necrotic and drop. New ones appear pinkish purple at first, and eventually become light yellow-green. Symptoms become most apparent during late summer.

3.2. Manganese (Mn) Deficiency

With this deficiency, younger leaves develop interveinal chlorosis, though the green coloration remains in a broader band than with iron deficiency (see 3.5). In fact, a manganese deficiency is often masked by an iron deficiency, since both elements become deficient under high soil pH. Leaf margins become necrotic, and the dead areas merge with necrotic spots on the blade.

3.3. Potassium (K) Deficiency

Initial symptoms appear as dead shoot tips, followed by dead spots on the leaf blades, and sometimes interveinal chlorosis on young leaves. In more severe cases, leaf margins scorch and plants are severely stunted. Leaves may be abnormally small and bronzed.

3.4. Magnesium (Mg) Deficiency

Symptoms usually appear first during harvest on older leaves near the lower part of the plant. Leaf margins turn bright red or yellow. Veins often remain green in a pattern resembling a miniature Christmas tree pointing toward the leaf tip.

3.5. Iron (Fe) Deficiency

Young leaves develop bright lemon-yellow, interveinal chlorosis, while veins remain deep green. Some leaves may develop a bronze hue and remain small. Older leaves remain very green. Shoot internodes appear normal. In severe cases, the entire leaf turns lemon-yellow, but is not cupped as in stunt. Necrosis associated with manganese deficiency (see 3.2) is not usually present.

3.6. Manganese Toxicity

Under very low soil pH manganese can become available in toxic quantities. Manganese toxicity mimics deficiency. Often a problem when sulfur is used to bring alkaline soils to a pH less than 5.

4.1. Rodent Damage

Vole damage is especially troublesome in plantations maintained in unmown sod or in such organic mulches as wood chips or straw. Voles may gnaw the roots or the bark on the crown or the canes near the mulch or soil line. Where such damage occurs, the plant may leaf out, but will subsequently die; in less severe cases, the entire plant may show drought and/or nutrient deficiency symptoms in summer. To reduce the incidence of such damage, keep the area near the bush clean through the use of herbicides or cultivation, or keep the sod mown short. The use of commercially available poisons is also effective.

Rabbits gnaw the bark higher up on the canes, as well as the tips and buds of low-hanging shoots. Shoot stubs have a clean edge compared with the ragged edge resulting from deer browsing. Where damage is severe, use one of the commercially available rabbit repellents.

Woodchucks or groundhogs sometimes claw and scratch the bark on blueberry canes, and their burrows can damage the root system.

4.2. Bird Damage

Some species eat the flower buds during the dormant season, removing them entirely from the plant. Many species will attack the ripening fruit, often pecking at several berries within the cluster without removing them. This is particularly a problem near wooded areas and during drought.

4.3. Nematodes

These microscopic, worm-like animals can cause substantial damage in some areas under certain conditions. Several species damage the root system, especially in cutting beds. The stubby root nematode attacks new roots emerging from callus tissue, resulting in poorly rooted or dead cuttings. The dagger nematode also feeds on roots and is the vector for necrotic ringspot.

4.4. Diseases

1. *Shoestring*. Early symptoms include red discoloration along the mid-vein, often extending part way up lateral veins. Leaves are wavy, distorted or crescent-shaped, or, in severe cases, narrow, pointe,d and light green to red in color. Shoots show red streaks along surfaces exposed to the sun. These shoots twist and break easily. Corollas may show red or pink streaks. Normally, green berries may be pinkish or purplish green. Note that 'Blueray' tissue usually exhibits some red coloration.

2. *Necrotic Ringspot*. Leaves are distorted and reduced in size and show chlorotic spots, rings, or lines. By looking through a leaf held toward the sun, its chlorotic areas will become visible. Shothole may develop, followed by considerable shoot dieback and stunting. Fruit is symptomless.

3. *Red Ringspot*. The symptoms are most noticeable in late summer. Red spots appear on the upper surface of leaves on lower and center parts of the bush. Red rings, spots, and raised areas may appear on the stem. Fruit may show red mottling while maturing.

4. *Stunt*. Symptoms include great vigor reduction, shortened internodes, and chlorosis. Early leaf symptoms are yellow tips and margins, interveinal chlorosis, and small, round, puckered, and cupped blades. In late summer, chlorotic areas turn a bright red, the color lying in two bands inside the leaf margin and parallel to the midrib. Berries are small, dark, and bitter; hard to separate from the bush; and may hang all winter. Plants in advanced stages of the disease will not produce. Red leaves caused by magnesium deficiency (see 3.4) appear on basal portions of shoots, while those caused by stunt appear near the tip of the shoot. The disease is more pronounced on plants that have been heavily pruned.

5. *Mosaic*. Early symptoms are light green or yellow leaf mottling. The color intensifies as the season progresses, with yellow turning to red. The coloration is more intense on lower leaves. Terminal leaves may show no symptoms. The whole bush will show symptoms after a few years. Fruit appear normal.

6. *Fusicoccum Canker*. The pathogen enters the shoot at a bud site or through injured tissue. The first symptom is some small, reddish coloration at the infection site in early spring. This is often near the soil line. This canker enlarges, with margins remaining reddish and the center of the spot turning gray or brown. The canker eventually girdles the stem, which flags, wilts, and dies suddenly in hot, dry weather. Leaves on infected canes will turn red prematurely in the fall.

7. *Cornyeum Canker*. This is sometimes a problem in the northeastern United States. The pathogen enters the cane through a wound, often the result of sunscald. Numerous saucer-shaped fruiting bodies appear in small sunken areas in the bark. The resulting canker girdles the cane.

8. *Botryosphaeria Canker*. This pathogen is widespread, though particularly devastating in the southern United States. It invades new shoots through the lenticels on the sunny side. Red swellings appear at that point of invasion by early fall. Black fruiting bodies later develop on these swellings. During the second growing season, the red coloration disappears, and deep fissures develop at the points of infection. Newly infected tissue swells and turns gray. The gray canker girdles the cane over the next few years.

9. *Bacterial Canker*. This is a major problem in the Pacific Northwest. In winter, water-soaked areas appear on one-year-old canes. These brown or blacken, and may extend the entire length of the stem. All buds in the infected areas die. Infection often occurs through injured tissue, and it can be confused with *Botrytis*, scorch, and frost damage.

10. *Witches' Broom*. Infection results in excessive lateral bud break in the spring. Infected branches appear swollen and cracked. There are no leaf or fruit symptoms. Fir (*Abies sp.*) is alternate host, and the pathogen must go from fir to blueberry to fir; it cannot go from blueberry to blueberry, or from fir to fir. Both plants must be within several hundred yards of each other for infection to occur.

11. *Crown Gall.* Bacteria enter through deep wounds in root or cane, and hard tumors or galls appear near the base of the cane. Canes wilt after being girdled with the gall. This is primarily a problem in nursery fields and young plantings.

12. *Anthracnose.* This is widespread throughout major blueberry areas of the eastern United States. The fungus is spread from previously infected tissue (in which it overwinters) to new tissue, by warm spring rains. It can cause stem cankers (which is rare), blossom cluster blight, and leaf spots, which vary from small spots to large dead areas. The earliest symptom of infection is shoot blight. The most definitive evidence of infection is small, black fruiting bodies covering the fruit surface. Under moist conditions, these "bloom" into masses of salmon-colored spores.

13. *Mummyberry.* The fungus overwinters in small, gray globes on the ground. With spring rains, cup-like structures form and discharge spores around the time of bud swell. Spores infect very young shoots and flowers, which blacken and wilt suddenly. They do not attack older tissues. Sweet-smelling, powdery fruiting bodies appear on infected tissue and are carried to open flowers by bees and wind. Ovaries become infected and developing seeds abort. Infected fruit appear to develop normally nearly to maturity, but instead of turning blue, they develop a salmon color and fall to the ground. These become the mummies in which the pathogen overwinters.

14. *Stem Blight.* Arkansas and North Carolina growers are particularly affected by this disease. The pathogen commonly attacks only one side of a stem. Infected wood turns pecan-brown and can extend the entire length of the stem. During the initial stage of infection, leaves become chlorotic or red. Later, the entire branch dies, while stems nearby remain healthy.

15. *Botrytis.* Spores are carried by spring rain and wind to new tissue. Cool, wet periods for up to a week often result in serious infection. Infected tips of new shoots turn black, then gray, and may resemble those damaged by severe winters, frost, and salt. Infected corollas turn brown, display abundant mycelia, and remain attached to the ovary. Poorly pollinated (see 1.5) and frost-damaged (see 1.4) flowers show no mycelia. Rapid, succulent growth common in overfertilized plantings is more susceptible to attack.

16. *Powdery Mildew.* This is widespread in the southeastern

United States and in Michigan. The upper or lower (or both) surfaces of leaves are covered with a white mycelium. Usually, the lower surface is attacked first. Red-bordered chlorotic spots appear on the upper surface. Directly beneath these, water-soaked areas appear on the lower surface later in the season. This differentiates the disease from red ringspot (see 4.4.3). By the end of the season, orange (later black) fruiting bodies form on the lower leaf surface. This disease weakens the plant over time.

17. *Phytophthora Root Rot.* This is a problem in wet soils. Leaves will be chlorotic or red, and drop prematurely. Shoot growth is reduced. Wilted branches are particularly noticeable in early morning. The normally white vascular tissue of the crown, root, and cane will be stained a reddish brown. Discoloration is extensive. Infected canes will be brown near the soil line.

18. *Phomopsis.* This disease becomes a problem on weakened canes or bushes. Young shoots are invaded first through glower buds, wilt suddenly, and become crooked. Shoot tips blacken, and damage is often mistaken for cold damage (see 1.3.4), drought (see 1.1), or boron deficiency. The fungus moves steadily into older tissue, where it can eventually girdle the cane. Red spots up to 0.5 in (1 cm) in diameter sometimes appear on leaves of infected shoots.

19. *Double Leaf Spot.* This can be a serious problem in North Carolina. Circular spots 0.12 in (2.3 mm) in diameter develop on leaves during wet springs. The spots are gray or light brown and are surrounded with a dark-brown ring. By midsummer, the spots become cinnamon-brown and enlarge to several millimeters in diameter. The original spot appears to be surrounded by a larger spot.

20. *Septoria Leaf Spot.* This is a problem in North Carolina and the southern United States. Small, round white spots with purple borders form on leaves and young shoots. Occasionally, the spots are tan with brown borders, or entirely brown. Heavy infection by any leaf spot interferes with photosynthesis.

4.5. Insects

1. *Borers.* Plants are affected individually throughout the planting, though initial damage often occurs in bushes near the perimeter.

A. *Black Vine Borer:* This is a problem especially in the Pacific Northwest. It deposits eggs in the soil near the plant crown, where feeding grubs girdle the cane or entire bush just below the soil surface. Affected plants are weakened and leaves redden prematurely.

B. *Blueberry Stem Borer:* This borer is a serious pest that is common in North Carolina and other areas along the east coast of the United States. Eggs are deposited between two parallel rings of punctures cut around the stem about 4 in (10 cm) from the shoot tip, which wilts and blackens suddenly. Yellowish larvae tunnel upward, killing the tip of the shoot; they then turn and bore downward 2-10 in (5-25 cm) during the first season. Frass is ejected through small holes chewed to the outside. By the end of the second year, larvae have bored to the base of the crown. By the third season, larvae have attacked several stems. Canes wilt and die.

C. *Blueberry Tip Borer:* Commonly, this small pink worm bores into the tips of new shoots in early summer, causing them to flag and purple. Their leaves turn yellow and develop red veins. Borer tunnels may extend nearly 12 in (30 cm) into a shoot. Future fruit production on this shoot is ruined.

2. *Root Weevils.* There are several root weevils that cause economically significant damage in the Pacific Northwest. Most important are the strawberry root weevil, the rough strawberry root weevil, and the black vine weevil. All are darkly colored; the black vine weevil has orange spots on its back. Larvae have brown heads, white bodies, and are bent into a "C" shape. Larvae begin feeding on roots in midsummer, and continue this activity throughout the winter. They pupate in midspring. Adults hatch during the late-bloom period and feed on leaves at night. Notching of the leaves is indicative of feeding adults. Eggs are laid in the soil in midsummer, and the cycle repeats.

3. *Japanese Beetles.* The adult will skeletonize the leaf and feed on the fruit. Its larvae can feed on blueberry roots, and they are a special problem under organic mulches.

4. *Bud Mite.* This is a problem in North Carolina, particularly following mild winters. In spring, mites migrate to areas beneath scales of newly forming buds; here, they mate, feed, and burrow toward the bud's center. Feeding and egg-laying continue into the winter. Injured plant parts show a roughened and red discoloration.

Sometimes, small red pimples appear on fruit. Buds fail to expand and produce fruit.

5. *Blueberry Maggot*. The adult fruit fly, about the size of a housefly, deposits a single egg beneath the skin of each fruit. Within a week, the egg hatches into a colorless larva that begins feeding immediately on the fruit pulp. As larvae mature, they turn white and, after about three weeks, drop to the ground to pupate. They remain buried in the soil for up to three years before hatching. Their peak time for emergence is just after the first berries begin to ripen, and egg-laying begins a few weeks after emergence. Infested fruit will be soft and leaky and, upon examination, may contain the maggot.

6. *Fruitworms*. There are two kinds of worms that cause major damage to the blueberry.

A. *Cranberry Fruitworm:* The grayish brown night-flying moth emerges from the soil when the first fruit begin to enlarge. The female deposits her eggs inside the rim of the calyx cup. In about a week, small green caterpillars hatch and enter the berry near the stem end. They enmesh several berries together, then eat the fruit. Most of the fruit interior is eaten and filled with frass.

B. *Cherry Fruitworm:* The dusk-flying, dark-gray moths with deep-brown wing bands lay greenish white, flattened eggs on the undersides of leaves during bloom. Other eggs are deposited later on green fruit. The white-bodied, black-headed larvae hatch in about a week, and slowly turn pink. They enter the fruit through the calyx and turn red a few days after beginning to feed. Entrance holes are sealed with silk, and the only visible signs of infection may be prematurely blue, shriveled fruit. They overwinter in a silk cocoon in debris beneath the bush.

7. *Plum Curculio*. The dark-brown, long-snouted adult is most active when temperatures are above 75°F (24°C). Females lay a single egg in immature green fruit, leaving a typical D-shaped puncture. After about a week, grubs hatch and feed on the inside of the fruit for about three weeks. Fruit shrivel and ripen prematurely, then drop. Look for adults at dawn and dusk by placing a white sheet beneath the bush and shaking the plant to dislodge the insects.

8. *Blueberry Blossom Weevil*. This is a major pest in New Jersey and in some New England States. The adult is a small, deep-red

snouted beetle that emerges during spring bud swell. It enters the clustered buds and bores small brown holes into the unopened blossoms to feed and lay eggs. Infested flowers turn purplish and drop unopened. The grubs eat the fallen ovary. Do not confuse this with unfertilized blossoms caused by poor pollination (see 1.5), in which the flower is fully opened, or with bumblebee damage (see 4.5.10).

9. *Putnam Scale.* This is a major pest in New Jersey, particularly in older plantings. Scales overwinter as adults, forming what appear as dull crusts on the wood surface. Crawlers emerge during bloom and move throughout the bush, sucking sap and excreting honeydew over the leaves and fruit. A black, sooty mold grows on the honeydew and interferes with photosynthesis and fruit quality. Plant vigor and yield decline. Scales congregate mostly under bark, where they appear as dull, waxy flecks. They are encircled by red rings on the fruit.

10. *Bumblebees.* These sometimes bore small holes at the base of the corolla, robbing the nectar and circumventing pollination. Honeybees may also use the holes to gather nectar.

11. *Blueberry Leaf Miner.* During their early life, the larvae mine in the lower leaf surface. They then emerge and roll the leaf tip back against the margin, tying it with silk to form a triangular tent. The larvae feed on the leaf surface beneath the tent.

Index